Processability of Polymeric Composites

Anup K. Ghosh
Mayank Dwivedi

Processability of Polymeric Composites

Springer

Anup K. Ghosh
Department of Materials
Science and Engineering
Indian Institute of Technology Delhi
New Delhi, India

Mayank Dwivedi
Directorate of Industry Interface
and Technology Management
Defence R&D Organisation Headquarters
New Delhi, India

ISBN 978-81-322-3931-4 ISBN 978-81-322-3933-8 (eBook)
https://doi.org/10.1007/978-81-322-3933-8

This Springer imprint is published by the registered company Springer Nature India Private Limited
The registered company address is: 7th Floor, Vijaya Building, 17 Barakhamba Road, New Delhi 110 001, India

Preface

Processing of advanced composites has witnessed a phenomenal growth over the last few decades. It is expected to grow at even a much faster rate due to the emergence of new processing technologies and the availability of a range of reinforcements and matrices. The choices offered by the variety of reinforcements and matrices pose a challenge to selecting the right combinations of materials—not only from the point of view of the desired properties of the end product but also from the processability. Both quality and cost-effectiveness are also addressed. The availability of various processing technologies and their combination makes this task further complex.

Processability is the capability of any system or material to be converted into the desired outcome or product. The processability in advanced composites is influenced by many factors, such as the chemical nature of the matrix and reinforcement, interactions between phases, physical dimension of the reinforcement, additives, surface treatments on reinforcement, rheology of the matrix or system, temperature, pressure, and so on. Processability of constituents of composites is the key to processing advanced composites for a desired outcome. In this pursuit, there will be a greater demand for trained and highly skilled manpower now than before. Keeping this demand in mind, the authors have written this book for the students, engineers, and technologists who are pursuing a career in the field of processing advanced composites.

This book covers various topics of interest with special references to distinct characteristics of processability in the processing of advanced composites. It starts with ► Chap. 1, which explains processability and its dimensions. This chapter further describes the classification of advanced composites, properties (physical mechanical, chemical, thermal, etc.) of matrices and reinforcements, and types of reinforcements (in terms of fibres, weaves, particulate, whiskers, etc.). It also discusses factors for quantitative understanding of processability.

In order to make our readers appreciate the potential of polymeric composites in various sectors, applications and advantages of polymeric composites are discussed in ► Chap. 2. This chapter gives an insight to the enormity of polymeric composites. It is stressed that the processability aspects for advanced composites have a direct impact on the performance of the product.

► Chapter 3 discusses micromechanics and macromechanics of advanced composites to impart a better understanding of composites and their mechanical behaviours under stress and strains. This chapter outlines theories for the basic understanding of the performance of advanced composites including the modes of failures. The influence of interfaces of a matrix and reinforcement for creating interphases in advanced composites and role of interphases in the composites are also emphasized in this chapter.

The rheology of a matrix and the rheology of a matrix under the influence of reinforcement are explained in ► Chap. 4. The important aspects of processability of thermoset matrices, thermoplastic matrices, and organic/inorganic/hybrid reinforcements are discussed. The selection methodology for matrices and reinforcement in regard to their processability and to develop desired composite product is explained. The differences between theoretical assessment and practical realities in processing, depending on the processability of advanced composite, are emphasized in this book. The processability of preforms and prepregs is also mentioned for a broader understanding of the readers in the area of advanced composite manufacturing. These discussions provide an assessment on the processability of advanced composites and the role of its constituents.

Polymeric composites and their processing technologies are discussed in ► Chaps. 5 and 6 to provide ample understanding on various properties, parameters of materials, and processing conditions that determine the processability of polymeric composites. ► Chapter 5 discusses the processability of thermosetting composites, whereas ► Chap. 6 discusses the processability of thermoplastic composites. Thermosetting composites are considered to be workhorse of composite industries. The role and functions of thermosetting matrices, interaction with reinforcement, and compatibility with processing technology have been elaborated. On similar lines, the roles, interactions, and compatible processing technologies of thermoplastic matrices have been discussed from the point of processability.

The processing of composites is an intensive study; therefore, the selection of processing technologies for making composite products plays an important role for the desired outcome and properties. The processability aspects of open mould processing technologies (hand layup, filament winding, tape winding, etc.) are deliberated in ► Chap. 7. Open mould technologies are generally adopted for large components and for costs to be controlled. Closed mould technologies have been elaborated in ► Chap. 8. Closed mould processing technologies (matched die moulding, compression moulding, autoclaving, injection moulding, extrusion, pultrusion, etc.) are used for production of composite products with high dimensional accuracies. The cost of the product depends upon processing technologies.

The efficacy of processability of advanced composites is vindicated by the quality of the resultant composite product. Therefore, ► Chap. 9 of this book discusses the characterization and testing of advanced composites with respect to the structure and morphology of advanced composites. This last chapter gives the complete picture of mechanical, thermal, and viscoelastic behaviours of polymeric composites. This will enable the reader to identify and correlate the processability aspects of polymeric composites, leading to the desired end product. ► Chapter 9 also discusses the structure–property relationship in composites.

This book is focused on the processability of advanced composites with respect to various processing technologies. It provides much-needed insight on the processability aspects suiting specific processing technologies. This book covers a broad horizon on various aspects of processability to enable the reader to assess and correlate advanced composites manufactured by different processing technologies. This will help students, engineers, and technologists to select right processing technology for processing specific advanced composites.

Anup K. Ghosh
New Delhi, India

Mayank Dwivedi
New Delhi, India

Contents

About the Authors

Anup K. Ghosh

Distinguished and Renowned Professor in the field of Polymer Science and Engineering, Indian Institute of Technology Delhi (IIT Delhi), India, and Fellow of the National Academy of Sciences, India

Ghosh obtained his PhD in Chemical Engineering from the State University of New York (SUNY), Buffalo, USA. He joined the IIT Delhi Faculty in 1991. Professor Ghosh's research interest includes rheology and the processing of polymers, reactive extrusion, polymer blending and alloying, polymer nanocomposites, cellular polymers/biopolymers, mixing polymers, and polymer reaction engineering. He has authored more than 200 journal and conference papers and edited two books, and is a co-inventor of nine patents. He is a recipient of the Young Alumnus Award of Chemical Engineering Department from Calcutta University, and the Meritorious Service Award from the Indian Plastics Institute, India. Professor Ghosh is a Member of many prestigious committees and boards, including the Editorial Boards of *Journal of Plastic Film & Sheeting* and Associate and the Editor of the *Journal of Packaging Technology and Research*. He is the founding member of the Polymer Processing Academy. He has been elected as the international representative of the Polymer Processing Society, USA, and is a Fellow of the Indian Plastics Institute.

Mayank Dwivedi

Senior Scientist and Director of the Directorate of Industry Interface & Technology Management (DIITM) in the Defence Research and Development Organisation (DRDO), India

Dr. Dwivedi completed his Masters of Engineering (ME) in Polymer Technology from the Delhi College of Engineering and Doctorate in Advanced Composites from the Indian Institute of Technology Delhi (IIT Delhi), India. He is an alumnus of the prestigious National Defence College (NC), India. He is member of many professional societies in the area of polymer engineering and allied fields. He is recipient of 'Lab Scientist of the Year' award in 2003 and 'Agni' Award for Excellence in Self Reliance in 2018. He has been working in the area of processing of advanced composites for more than 30 years, and his work includes the development of airframes, high-temperature composites, polymeric nanocomposites, and biomechanical devices. He has a number of international publications, patents, and conference papers to his credit. He has authored two books and edited numerous chapters in the area of advanced composites and nanotechnology.

Introduction

The original version of this chapter was revised with the late correction received from the author. The correction to this chapter can be found at https://doi.org/10.1007/978-81-322-3933-8_10

A. K. Ghosh, M. Dwivedi, *Processability of Polymeric Composites*,
https://doi.org/10.1007/978-81-322-3933-8_1

Composites have existed in nature right from the evolution of life on Earth. Wood is the finest example of natural composite in which cellulose is reinforcement and lignin is matrix. The combination of bones and musculature is another example of a natural composite where bones are reinforcements and muscles are the matrix. Similarly, in orthopods, the exoskeletal body is reinforcement, and interior soft tissues are matrix. By definition, composites are the materials/structures that have two or more physically distinct constituent phases, and the resultant properties of these materials/structures are superior to the properties of its constituent phases. These phases are primarily reinforcement and matrix and are separated by a distinct interphase. Interphase has a different combination of properties from those of constituent phases. In hybrid composites, different types of reinforcements are used. A composite material also may be defined as the heterogeneous material that consists of two or more constituent materials which do not lose their characteristics. This combination of constituent materials results in new desirable properties. Normally, the constituents (reinforcement and matrix) can be physically identified along with the interface, which also generally controls the properties of the composites. The constituents can be organic, inorganic, or metallic in the form of particles, rods, fibres, plates, foams, etc., and they do not dissolve or merge completely into each other [1, 2].

The essence of science in composites has been understood from time immemorial since straw- or bamboo-reinforced mud houses were built for the purpose of dwelling. The first use of composites dates back to around 3400 B.C. by the ancient Mesopotamians when they glued wood strips at different angles to create plywood. Egyptians and Mesopotamian settlers used a mixture of mud and straw to create strong and durable buildings in fifteenth century B.C. Straw continued to provide reinforcement to ancient composite products, including pottery and boats. Houses and kilns made out of straw-reinforced mud also were used by Indians from ancient times. Harappa and Mohenjo-daro have plenty of evidence of reinforced structures that were none other than composites. The use of *wattle* and *daub* as composite building material has a history of 6000 years. In these composite structures, a woven lattice of wooden strips called *wattle* is used as reinforcement, and the matrix is a combination of wet soil, clay, sand, animal dung, and straw (called *daub*). *Wattle* and *daub* are still important construction materials in many parts of the world [3]. For modern engineering applications, synthetic polymeric matrices (such as epoxy resin, polyester resin, polyimide resin, nylon, etc.) are used with naturally occurring mineral fillers (such as wollastonite, silica, mica, talc, and calcium carbonate) and synthetic fibres (like glass, carbon, and Kevlar®). Any solid material that can be processed so that it can embed and adherently grip a reinforcing phase (i.e. fibres or fillers) is a potential matrix material for composites.

This engineering practice of the fabrication of composites using natural materials got its boost when man-made fibres and matrix materials were synthesized. The invention of Bakelite in 1895 was the stepping stone. In the early 1900s, plastics (such as polyvinyl alcohol, polystyrene, phenolic, and polyester) were developed. These new synthetic polymers outperformed natural resins. Thereafter, it was the turn of newer reinforcing materials. The first glass fibre was developed in 1935 (from Owens Corning). Fibre glass, when combined with a polymer, created an incredibly strong structure that was also lightweight. The first composite boat was made by Ray Greene of Owens Corning in 1937, but the work did not progress further due to the brittle nature of the matrix used. In 1939, construction of a passenger boat of plastic materials was reported from Russia, and similarly, the fuselage and wings of an aircraft were reported from the United States. In 1946, the first car, the Stout Scarab, having a fibre glass body was built [4, 5]. This was the beginning of the fibre-reinforced polymer (FRP) industry as we know it today. This development was further fueled by World War II with the first high-performance application in the form of *radome (radar + dome)* made out of glass fibre-reinforced polyester resin for fighter aircrafts of the Royal Air Force in 1940.

In India, advanced composites were first used by Dr. APJ Abdul Kalam to make the nozzle of the satellite launch vehicle (SLV-3) in 1980. He pioneered the research in the area of advanced composites and can be truly referred to as the 'father of advanced composites' in India. Similarly, Professor KAV Pandalai from IIT Madras was the first teacher of advanced composites in India. He started his teaching career from the Madras Institute of Technology (MIT), and he even taught aero structures and their analysis to Dr. Kalam in MIT [6].

Point to Ponder...
Two of the famous quotes of Dr. Kalam are 'Strength respects strength' and 'If you want to shine like a sun, first, burn like a sun'.

A unique feature of composites is that the properties can be tailored to a specific engineering requirement using a careful selection of matrix, the reinforcement type, and the direction of it fibres. Therefore, a composite product is designed in such a way that the fibres are laid in the direction of maximum stress so that its performance under load is good. Generally, a fibrous phase (reinforcement) is embedded into a continuous phase (matrix). Polymer matrix composites (PMCs) or FRPs constitute a major category of composite materials with a wide range of applications. Anisotropy is the main characteristic of composites. The deformation of a composite is nonuniform/unsymmetrical on the application of a uniaxial load due to the anisotropic behaviour of composites, whereas the deformation in isotropic materials is symmetrical on the application of a uniaxial load.

The growth of advanced composites has spurred due to availability of various choices of raw material, processing technologies, and their adaptability to get processed into any intricate shape. However, striking the right combination of raw materials and processing technology to obtain the desired cost-effective composite product is governed by its processability.

1.1 Processability

Processability is the capability to transform the given input into desired output. The input may be in terms of data, signal, material, parameter, etc., and the output may be a document, table, actuation, digital display, product, etc. The processability in polymeric composites is defined as a set of input in terms of reinforcement and matrix that are converted into a desirable polymeric composite product through appropriate processing equipment and using optimized processing parameters. The term desirable is important because it qualifies the technology component in processability or, in other words, qualifies for technological matured preprocess where quality, cost, and customer satisfaction are catered for. Processability defines the processing of a particular composite that is amenable to a particular processing method or not. For example, processability of short, glass fibre-reinforced nylon composite using injection mouldings is possible to fabricate a product, but the same product is not processable using long glass fibres or epoxy resin in the injection moulding process.

Example 1.1 Why are glass or carbon fabric-reinforced nylon composites not processed through a direct injection moulding process?

Answer

There is a limitation of processability of glass/carbon fabric in direct injection moulding due to the restriction of the physical dimension of reinforcement that can't pass through the injection barrel and nozzle of injection moulding machine. The fabric architecture does not allow the composite mixture to flow through the injection barrel and the injection nozzle; therefore, direct injection moulding is not feasible.

However, it is to be noted that in automated processing, such as injection moulding, the processing is governed by the allowable processing limits offered by the moulding technology. However, cases of semiautomated or manual processing technologies (such as hand layup, compression moulding, etc.) allow a higher freedom of choices of raw materials, creating wider processability options.

Processability provides a window to produce desired products with one or many processing options. The correct processing option will ensure quality, cost effectiveness of the product, and customer satisfaction. The main purpose of processability is to ensure functionality, quality, and cost of the product to suit the customer demand. Therefore, processability directly controls the properties, performance, and cost of the product, as shown in Fig. 1.1.

Processability delves in both choice of materials and processing technologies in the design and manufacture of polymeric composites. It offers choice of processing technologies according to the application of desired product. For example, composite products with large dimension tolerances may be made using a contact layup process, whereas the same product may be made using a compression moulding process with higher dimension accuracies. The difference in the cost of manufacture in terms of cost of processing equipment, mould/die, choice of materials, expertise, maintenance costs, etc. is much higher for the compression moulding process than for the contact layup process.

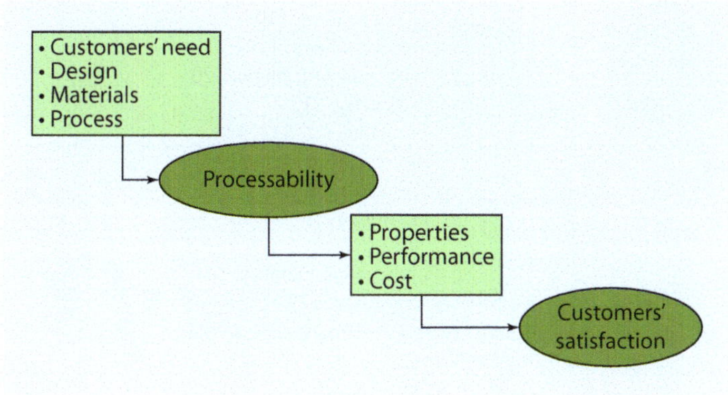

Fig. 1.1 Processability leading to customer satisfaction

Table 1.1 Compatibility factor between reinforcement and matrix

Serial no.	Reinforcement	Matrix	C_f
1.	Hydrophilic (metallic powder, silicon carbide, untreated glass fibres, etc.)	Hydrophobic (epoxy, polyester, nylon, etc.)	0.1
2.	Hydrophobic (silane-treated glass fibres, Kevlar fibres, sized carbon fibres, etc.)	Hydrophobic (epoxy, polyester, nylon, etc.)	1.0
3.	Hydrophilic (metallic powder, silicon carbide, untreated glass fibres, etc.)	Hydrophilic (polyvinyl alcohol, starch, etc.)	0.2

The polymeric composites, having no residual or internal stresses, have very low or negligible configurational entropy. The processability constant (M) is the mean value of the compatibility factor between reinforcement and matrix and the compatibility factor of processing technology with materials. It may be written as

$$M = \frac{C_f + P_f}{2} \quad (1.1)$$

where
- M = Processability constant
- C_f = Compatibility factor between reinforcement and matrix
- P_f = Compatibility factor of processing technology with materials

The compatibility factors between reinforcement and matrix are shown in Table 1.1.

The compatibility factors of processing technology with materials are tabulated in Table 1.2.

1

Table 1.2 Compatibility factors of processing technology with materials

Serial no.	Composite system	Processing technology	P_f
1.	Thermosetting resins and short fibres/staples	Casting, layup, etc.	0.5
2.	Thermosetting resins and continuous fibres	All process except injection moulding, extrusion, etc.	1.0
3.	Thermoplastic resins and short fibres	Injection moulding and extrusion	1.0
4.	Thermoplastic resins and continuous fibres	All process except injection moulding, extrusion, resin transfer moulding, etc.	0.7

Example 1.2 A chair is made out of an E-glass CSM-reinforced polyester composite. The chair does not have any internal or residual stress. Calculate the processability constant for this chair. (Use compatibility factors from Tables 1.1 and 1.2.)

Answer

$$M = \frac{C_f + P_f}{2}$$

The values of C_f and P_f are1.0 and 0.5, respectively.

$$M = (1.0 + 0.5)/2 = 0.75$$

1.2 Dimensions of Processability

Point to Ponder...
Design is an iterative process that caters to the need of the end user. No design is perfect; thus it always can be improved.

As has been discussed, processability needs to govern quality, cost, and customer satisfaction in order to ensure the commercial viability of a product being produced, as shown in Fig. 1.2. In this case, the selection of materials and their processability for the most efficient processing technology and equipment become vital. This controls the overall economy, which finally drives the market and attracts the customer towards the product. It is important to appreciate here that not only the processability of material with particular processing technique is important, but the economy of processability offered by the processing technology adopted also plays a vital role. Therefore, the processability of advanced composites starts from the product design stage. Design is based on the need of the customer; therefore, requirements of the customer and the operational needs of the product are catered in the design. This calls for a

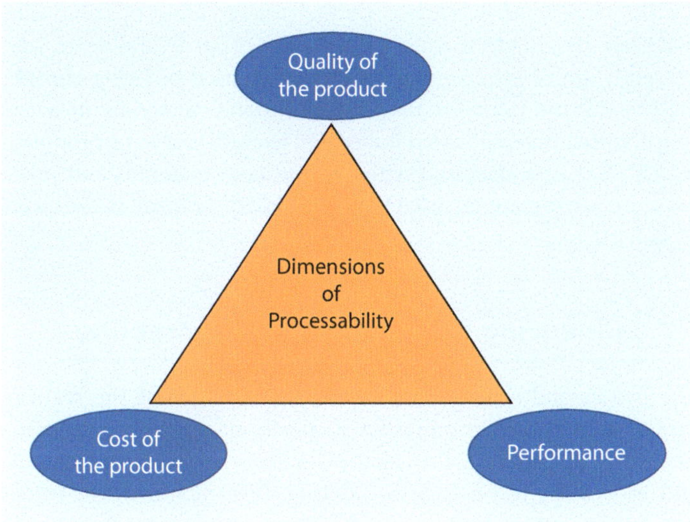

Fig. 1.2 Dimensions of processability

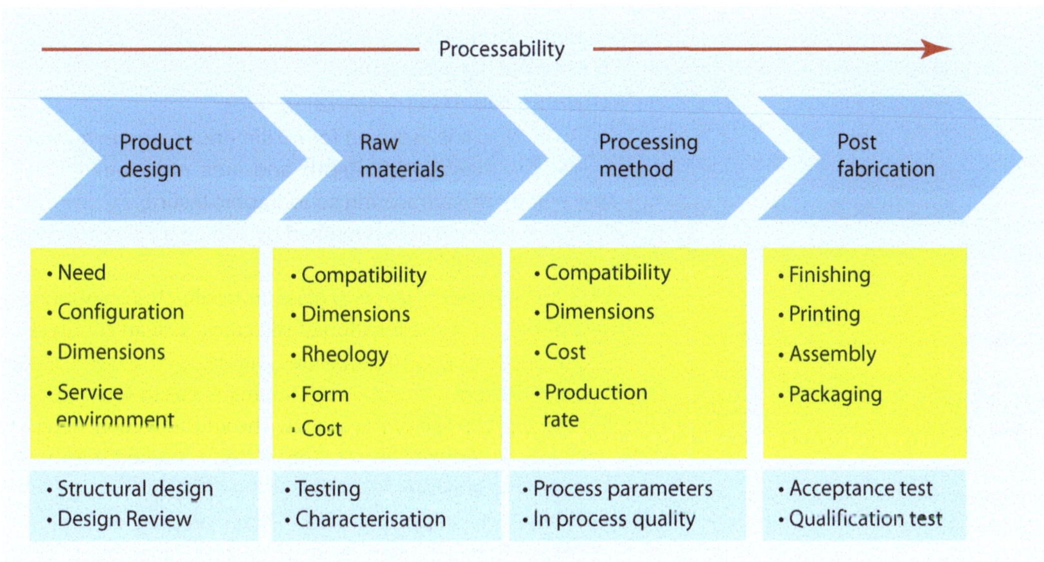

Fig. 1.3 Dependence of product development on processability

design that can be realized using polymeric composites. The confidence of realizability of design is the capability to convert it into a product. The next step is the selection processing of raw materials to realize the product. This means the processability of materials is governed by compatibility between reinforcement and matrix, percolation of resin, cost of raw materials, etc. Even during the processing of the polymeric composites, processing parameters may influence the processability and the final outcome as a product. Finishing, printing, or assembly of the product are also influenced by the further processability offered by the product. Therefore, the whole issue of processability is interlinked, and product development is dependent on processability, as shown in Fig. 1.3.

The large window created by the processability of composite materials has made them a major class of structural materials. Presently, polymeric composites are either used or being considered as substitutions for metals/conventional materials in aerospace, automotive, and other industries. Examples of some products, which were manufactured with conventional materials and presently are being manufactured using advanced composites, are given in Table 1.3.

1.3 Influencing Parameters in Processability

The processability in polymeric composites ensures the realization of a composite product for its performance with cost-effectiveness. The processability factors for the matrix, reinforcement, and processing technology will decide the outcome of the product. A brief picture of the processability parameters for the matrix, reinforcement, and processing of composites is given here. The detailed discussion on these aspects will be carried out in subsequent chapters.

1. *Processability Parameters of a Matrix*
   - Viscosity during processing
   - Wettability with reinforcement
   - Chemical interaction with reinforcement
   - Reactivity or evolution of byproduct
   - Cross-link density
   - Gelation time
2. *Processability Parameters for Reinforcement*
   - Surface treatment
   - Interphase
   - Thermal stability
   - Pliability/shaping
   - Percolation of matrix
3. *Processability Parameters for the Processing of Composites*
   - Compatibility with the form of reinforcement
   - Flow of matrix
   - Process compatibility of matrix
   - Production rate
   - Cost of production

These factors indicate that processability has a big influence on composite products as outlined here.

- *Performance of Composites*—Processability drives the performance of the composite. The selection of raw materials and processing technology yield a useable product with the required performance. If the compatibility of raw materials with the process is not there, the performance of the product is likely to suffer. Also, the design of the product dictates the selection of raw materials and processing technology. For example, a product having a shape undercut may not be processed by compression moulding technology [9, 10].

Table 1.3 Conventional materials replaced by polymeric composites [7, 8]

Serial no.	Product	Polymeric composites	Earlier materials
1.	Boat hulls 	Glass fibre-reinforced epoxy or polyester composite	Steel or wood
2.	Airframes 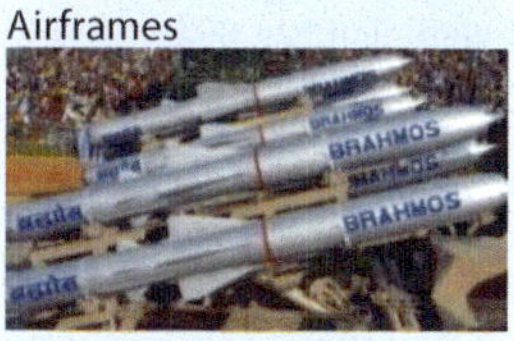	Glass or carbon fibre- reinforced epoxy/polyimide composite	Aluminium/ titanium/ maraging steel
3.	Front and back spoiler of automobile 	Glass fibre-reinforced polyester/polypropylene composite	Steel
4.	Fuselage of fighter aircrafts 	Carbon fibre-reinforced epoxy composite	Aluminium
5.	Artificial limbs and Calipers 	Glass or carbon fibre- reinforced epoxy/polyester/polypropylene composites	Aluminium/ steel/titanium
6.	Bulletproof panels 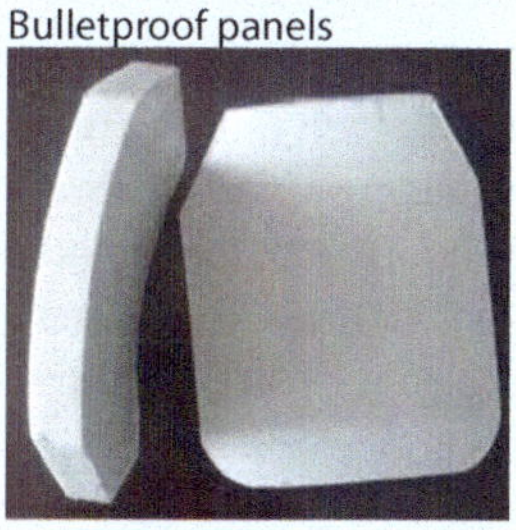	Ultra-high-molecular-weight polyethylene (UHMWPE) (UHMWPE) fibre-reinforced composites or Kevlar- reinforced epoxy composite	Ballistic grade steel

- *Energy Requirement*—The type of processing technology adopted for the fabrication of composites is the first requirement. For example, optimum heating and least post-fabrication operations reduce the energy requirements of the fabrication of the product. Energy saving also is linked with product design, selection of raw materials, process selection, and post-fabrication operations.
- *Cost-Effectiveness*—The selection of raw materials and maintaining the processing parameters make the composite product cost-effective. For example, a storage container may be made out of hand with a layup process using a split cavity mould and glass fibre-reinforced polyester resin. Another way is to make it using a filament winding process with a glass fibre-reinforced epoxy resin. The latter will cost more, whereas the first process will create the product for its intended use at comparatively extremely low cost.

1.4 Classification

Polymer-based composite materials are generally classified on the basis of the geometry of a representative unit of reinforcement. The reinforcement may be in the form of fibres or particles. *Particulate-filled composites* consist of a continuous matrix phase, while the *discontinuous filler phase* is made up of distinct particles. Fillers are finely divided solids that are originally introduced into polymers to lower the cost and to increase stiffness and hardness. Fibre-filled composites consist of a two-dimensional distribution of randomly oriented chopped fibres as the discontinuous phase (e.g. glass fibre-filled composites).

Generally, composites having a fibre length < 15 mm are called *short fibres*, and reinforced composites and composites having a fibre length > 15 mm are called *long* or *continuous fibre-reinforced composites*. The short fibres having aspect ratio (length/diameter) >100 are classified as *reinforcing fibres* [11]. A detailed classification of polymeric composites is given in Fig. 1.4.

Example 1.3 At what aspect ratio will short glass fibre-reinforced epoxy composites attain the maximum tensile strength?

Answer

In a well-aligned short fibre-reinforced epoxy composite, the strength of the composite will reach the maximum at $10l_c$ (l_c is the critical length of the fibre). An l_c of glass fibre is approximately 1.5 mm. Therefore, a 15 mm length of reinforcing glass fibre would be required to achieve maximum properties of the composite. $4l_c$ will yield up to 90% of the maximum composite strength.

Apart from previous classification, composites are classified based on the matrix material used in them.

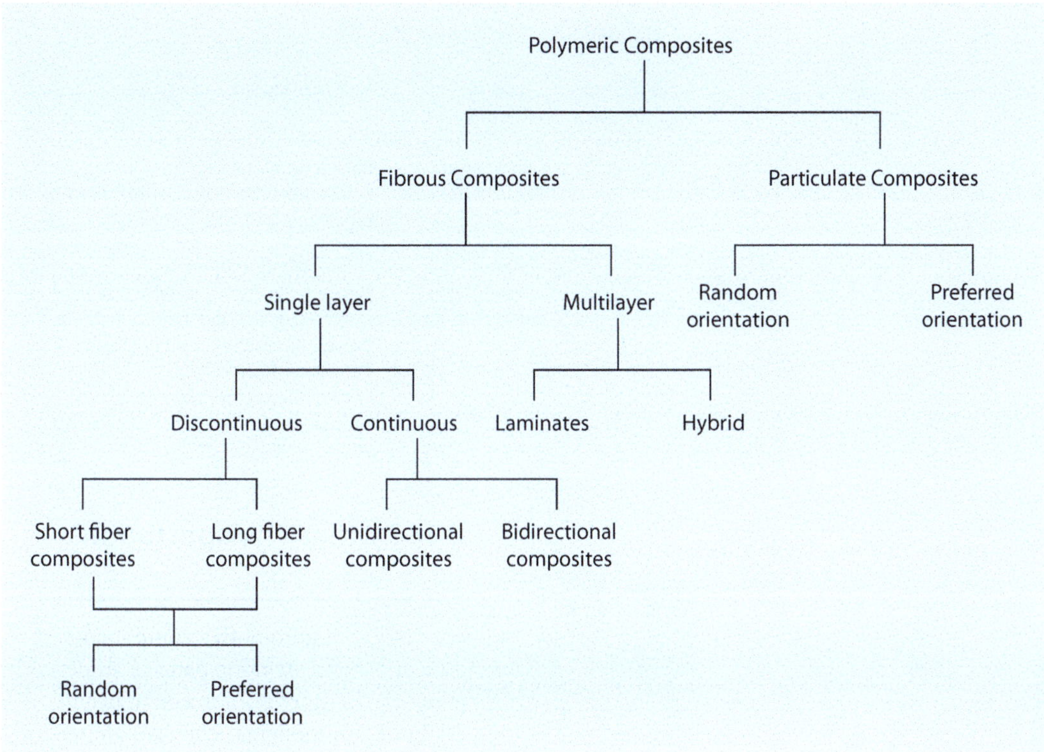

Fig. 1.4 Classification of polymeric composites

1.4.1 Polymeric Matrix Composites (PMCs)

PMCs are established engineering materials and are being widely used. PMCs are composed of reinforcement and matrix. Matrix is the continuous phase, and reinforcement is the dispersed phase. The matrix is classified into non-biodegradable and biodegradable (green matrices), as given in Table 1.4.

The reinforcement is classified into natural fibres and synthetic fibres. There is worldwide interest in green composites using natural reinforcing fibres and biodegradable matrices. Natural fibres offer a wide range of choices. The diameters of natural fibres are much higher (jute: 25–200 μm, flax: 40–600 μm, coir: 100–450 μm, sisal: 50–200 μm, pineapple leaf fibre (PALF): 20–80 μm, etc.) than synthetice fibres (carbon fibre: 5–8 μm, glass fibre: 8–15 μm, Kevlar fibre: 8–15 μm, basalt fibre: 9–13 μm, etc.). The classification of natural fibres and synthetic fibres is provided in Table 1.5.

The dispersed phase is also known as reinforcing phase, and it is available in the form of filament, yarn, rovings, tows, fabric, whiskers, tapes, etc. Detailed discussion on PMCs and their processing are described in the following chapters.

1.4.2 Ceramic Matrix Composites (CMCs)

CMCs are composed of a ceramic matrix using a continuous phase and embedded fibres of other ceramic material as a dis-

1

Table 1.4 Classification of matrix

Serial no.	Type of matrices	Matrix materials
1.	Biodegradable matrices	Polylactic acid (PLA), Polycaprolactame (PCL), polyhydroxybutyrate (PHB), etc.
2.	Non-biodegradable matrices	*Thermoplastic matrices* Nylon, polycarbonate, polypropylene, polyurethane, polyetherimide (PEI), polyether ether ketone (PEEK), etc. *Thermosetting matrices* Epoxy, phenol formaldehyde, bismaleimide, unsaturated polyester, etc.

Table 1.5 Classification of fibres

Serial no.	Type of matrices	Matrix materials
1.	Natural fibres	Jute, hemp, sisal, cotton, coir, bamboo, sunn, flax, kenaf, banana leaf, etc.
2.	Synthetic fibres	*Low-melting (thermoplastic) fibres* Nylon, UHMWPE, polypropylene, PEEK, polyester, polyurethane (Lycra®), etc. *High-melting fibres* Glass fibre, carbon fibre, Kevlar fibres, basalt fibres, etc.

persed phase. Ceramic matrix composites improve the fracture toughness of ceramics by reinforcing soft ceramic fibres in a hard and brittle ceramic matrix. In CMC, carbon, alumina, silicon carbide, etc. are reinforced in glass, glass ceramic, oxide ceramic, and other matrices. Among various techniques for their processing, slurry infiltration followed by sintering is the most common processing method for CMCs. In this technique, fibre is drawn through a slurry of ceramic, thereby coating the fibre with ceramic. This ceramic-coated fibre is dried to make a ceramic-coated fibre prepreg. This prepreg is placed in the mould and sintered at a high temperature to get CMC. The prepregging of fibres is easily processable; however, the prepregging of fabrics is difficult for CMCs. Other technologies for making CMCs are the sol–gel process, in situ chemical reactions, polymer pyrolysis (which yields carbon–carbon composites), etc. Examples of these are silicon carbide fibre-reinforced silica composites, boron nitride fibre-reinforced silica composites, etc.

1.4.3 Metal Matrix Composites (MMCs)

MMCs are composed of a metallic matrix (such as aluminium, magnesium, and beryllium) using a continuous phase with high-melting temperature metals (such as tungsten and molybdenum) or of non-oxide ceramics (such as silicon carbides, carbon, boron, and alumina) using a dispersed phase. The reinforcing fibres with ionic bonds have lower surface energies than the metals to be used as matrix. The incompatibility in surface energy does not allow the reinforcing fibres to make bonds with the metal matrix. To overcome this, the ceramic fibres are coated with a thin film of high-melting temperature metal that has a higher melting point that that of the metal matrix. The MMCs are fabricated by many methods, and the following are the most common:

- Alternate stacking of reinforcing fibres (SiCs) and metal matrix films (Al foil) are placed in the mould, and with the application of heat and pressure, the MMC is formed.
- Powder metallurgy technology uses a high-temperature isostatic press to apply heat and pressure to form the MMC. It uses metal matrix in the form of fine powder and reinforcement in the form of short fibre or whiskers.
- Plasma spray or chemical vapour deposition (CVD) of metal matrix over the reinforcing fibres yields a prepreg followed by pressing this prepreg in an isostatic press to yield MMC.
- Co extrusion of reinforcing fibres with a metal matrix yields towpregs. Fabrication of MMC is carried out by placing a towpreg in the mould and pressing it in isostatic press.

1.4.4 Carbon–Carbon Composites (CCCs)

CCC is a class of CMC. Although the initial processing of CCCs is similar to PMC, there is no polymer present in the final composite. It has carbon fibre as reinforcement and carbon as matrix. The common process of making CCC is by making a PMC having a high char yield containing a polymeric matrix such phenol formaldehyde. This prepared PMC is pyrolyzed at a high temperature followed by carbonization and graphitization. Graphitization takes place above 2700 °C. After every heat cycle, the polymeric resin is infiltrated to fill the voids caused due to charring in order to have good densification. After the last heating cycle, no polymeric resin infiltration is carried out.

Point to Ponder...
Ablation is a process in which an ablative material provides thermal protection from high temperatures by sacrificing its mass under thermal load. The temperature in a rocket engine is in excess of 2500 °C whereas, ablative liners keep the temperature < 50 °C at the outside surface across the thickness of approximately 50 mm.

Example 1.4 What makes the processability of CCC possible starting with raw materials, such as pitch-based carbon fibres and phenolic resin?

Answer
In CCC, the matrix and reinforcement are both carbon. Therefore, it is essential to have a matrix that yields into high carbonous content (i.e. charring yield). Special grades of

phenolic resins and epoxy resins have this property. Similarly, there is a need for carbon fibres with high thermal stability. Pitch- or rayon-based carbon fibres have high thermal stability. High charring content of resin causes high residual carbon mass in the composite; therefore, the number of cycles for saturating the composite with phenolic resin for densification of the matrix is reduced. This makes the process time short. Nose cones of hypersonic rockets and reentry vehicles are made out of CCC.

Polymeric composites are the most popular among all the composites. They use glass fibres, carbon fibres, Kevlar fibres, polyolefin fibres, natural fibres, etc. as reinforcements and a host of polymers (such as epoxy resin, polyester resin, polyamides, polyimides, polypropylene, etc.) as a matrix. Reinforced thermoplastics are the most rapidly growing class of composites. In this class, the focus is on improving the properties of the resin to allow these materials to perform well in new applications where conventional materials were used previously. Polymeric composites have versatility in terms of hybrid construction, wherein different types of reinforcements in the composite result in a range of properties that cater to the different needs of end-use applications. Today, polymer matrix composites find applications in every walk of life (such as aerospace, automotives, locomotives, marine, sports goods, household appliances, construction, oil and gas, electronics, etc.).

It is evident from this discussion that the two key components of an advanced polymeric composite are resin matrix and reinforcement. The types of resin matrices are thermosetting and thermoplastic matrices. *Thermosetting matrices* are epoxy resin, unsaturated polyester resin, phenolic resin, vinyl ester resin, polyimide resin, etc. They are solidified after the application of heat and are not deformable after curing as they form a 3-D cross-link structure. The most widely used resin matrix for advanced composites is thermosetting epoxy resin. Other thermosetting resins, such as bismaleimides and polyimides, are finding increasing applications in high-temperature environments. *Thermoplastic matrices* may be semi-crystalline, crystalline, or amorphous in morphology. In the former type, the semi-crystalline or crystalline morphology may be significantly influenced by the reinforcement, which may or may not act as nucleating agent. Today, the use of thermoplastic resins is comparatively much less than that of thermosetting resins in advanced composites. However, inherent toughness and potential recycling capabilities of thermoplastics have increased industry's interest in these resins. Some of the major thermoplastic resins which are being evaluated for advanced polymer composites include polyamide-imides (PAl), polyether ether ketone (PEEK), polyetherimides (PEI), polyether sulfones (PES), and polyphenylene sulphides (PPS).

Fillers show reinforcement characteristics depending upon the aspect ratio of the filler and the interphase interaction. In many cases, the treatment of fillers with coupling agents results into a chemical bonding between the filler and the polymer matrix. This chemical bond significantly improves the processability of filler in the polymeric composite. The term *reinforced polymer* is often used for composites with fibrous or reinforcing materials, while *filled polymers* are the materials which consist of particulate fillers. A vast number of filled polymers are used today. The most common filler for polymers are calcium carbonate, talc, and silica.

1.5 Properties of Reinforcements

The mechanical properties of composites depend on many factors, such as the material of the fibre, the fibres' orientations, and the fibre architecture. The properties of high-temperature reinforcing fibres and thermoplastic fibres are provided in Table 1.6.

The properties of natural fibres and select conventional materials are presented in Table 1.7.

Woven fabrics have led to easier handling and layup of composite laminates. In addition, composites made out of woven fabric are

Table 1.6 Properties of reinforcing fibres [9–12]

Serial no.	Material	Density (g/cm^3)	Tensile strength (MPa)	Tensile modulus (GPa)	Elongation at break (%)
	High-temperature fibres				
1.	E- glass	2.54	3500	72.4	3
2.	S- glass	2.48	4600	85.5	2
3.	R-glass	2.58	4400	84.8	2
4.	Carbon (high strength)	1.8	4000	390	1–2
5.	Carbon (high modulus)	1.9	2100	390	1
6.	Boron	2.63	3500	415	1
7.	Aramid (Kevlar 49)	1.44	3500	125	1.2
8.	Aramid (Kevlar 149)	1.47	3400	185	3
9.	Silica	2.19	5800	72.4	3
10.	Silicon carbide	3.18	2100	480	2
11.	Silicon nitride	3.2	410	310	2
12.	Alumina	3.7	1700	370	2
13.	Basalt	2.7	4840	89	3.15
	Thermoplastic fibres				
14.	Nylon	1.44	65	3	52
15.	Peek	1.27	90	4	60
16.	Polyester	1.5	45	3	50
17.	UHMWPE	0.94	1500	20	3
18.	Polypropylene	0.96	200	6	70

1

Table 1.7 Properties of natural fibres and conventional materials [9–13]

Serial no.	Material	Density (g/cm^3)	Tensile strength (MPa)	Tensile modulus (GPa)	Elongation at break (%)
	Natural fibres				
1.	Jute	1.3–1.45	393–773	13–26	1.1–1.5
2.	Hemp	1.47	690	70	1.6
3.	Cotton	1.55	287–800	5.5–12.6	7–8
4.	Coir	1.15	131–175	4–6	15–40
5.	Flax	1.5	345–1100	27.6	2.7–3.2
6.	Ramie	1.5	400–938	61–128	1.2–3.8
7.	Sisal	1.45	468–640	9.4–22	3–7
	Conventional materials				
8.	Aluminium alloy (6061)	2.7	140	70	12
9.	Aluminium alloy (2014-T6)	2.8	460	72	10
10.	Titanium (Ti-6Al-4 V)	4.5	930	110	18
11.	Steel (4340)	7.8	990	207	12
12.	Mild steel	7.8	450	200	15
13.	Tungsten	19.3	5500	360	5

tolerant and thus are less susceptible to damage. Woven fabrics are produced by the interlacing of warp (0°) fibres and weft (90°) yarns in a regular pattern or weave style. The fabric's integrity is maintained by the mechanical interlocking of the fibres. Fabric is made of yarns that have an inherent twist. The *twists* are of two types: 'S' and 'Z'. *Ply* is a strand of twisted fibres, and two, three, or more plies form a yarn. 'S'- type twist has comparatively loose architecture compared to a 'Z' type of twist in the yarn. The percolation of a matrix is easier in an 'S' type of twist when compared to 'Z' type of twist in the yarn [14–16]. *Drape* is the ability of a fabric to adapt or conform to a complex surface. The smoothness of surface and stability of a fabric are controlled primarily by the weave style. The description of some of commonly found weave patterns is given here.

- *Plain Weave*—The plain weave is the simplest of the weaves. In this, each warp yarn passes alternately under and over each weft yarn. The plain weave fabric is symmetrical and has good stability. However, its drapeability is poorest among all weaves. The high level of fibre crimps imparts relatively low mechanical properties when compared with the other weave styles. Because this weave pattern gives excessive crimp, it is not used for high-tex yarns. The plain weave pattern is shown in Fig. 1.5a.

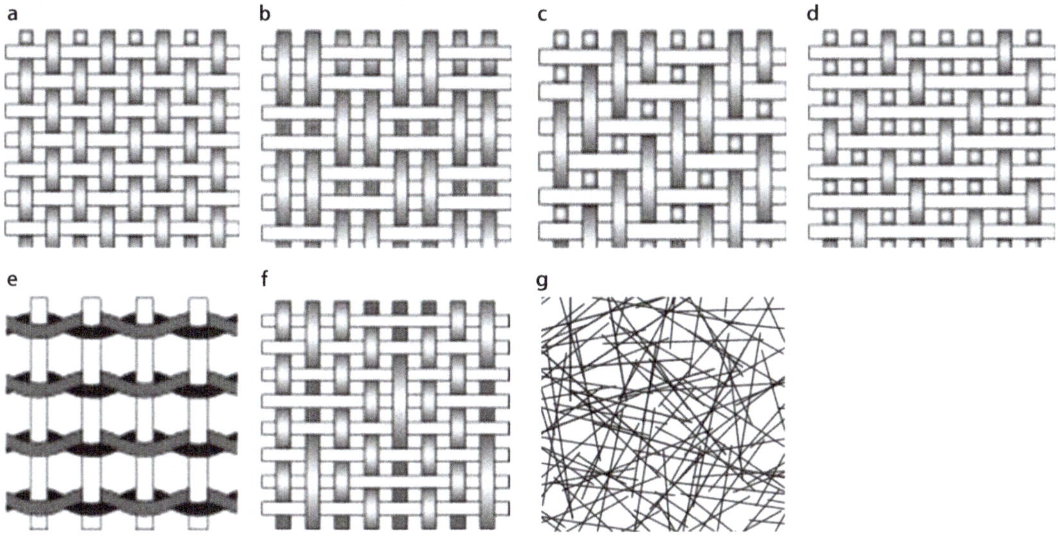

Fig. 1.5 Schematic representation of **a** plain weave, **b** basket weave, **c** twill weave, **d** satin (in this case a 4 harness) weave, **e** leno weave, **f** mock leno weave, and **g** CSM [14, 15]

- *Basket Weave*—A basket weave resembles the plain weave except that two or more warp yarns alternately interlace with two or more weft yarns (unlike in plain weave where only one yarn interlaces). A weave of two warp yarns crossing two weft yarns is called a 2 × 2 basket. The arrangement of fibre need not be symmetrical; therefore, the patterns of a basket weave may be 4 × 2, 4 × 4, etc. The texture of a basket weave is somewhat flatter, and it is less crimped than a plain weave due to interlace distance being increased by two or more warp/weft yarns. It is stronger than a plain weave, but it is less stable. The basket weave pattern is shown in Fig. 1.5b.
- *Twill Weave*—In a twill weave, one or more warp yarns alternately interlace over and under two or more weft yarns in a repeated manner. This imparts a visual effect of a straight or broken diagonal 'rib' to the fabric. This is the most used fabric pattern for reinforcement in the composite industry. It has good drape and more porosity than a plain weave; therefore, its wettability by resin is good. Comparatively, it is a little inferior in dimensional stability than a plain weave. Due to the repetition of warp yarn after long gaps (two or more weft yarns) , its crimp is less in comparison to plain weave. Therefore, the appearance of fabric is smoother with good mechanical properties. The twill weave pattern is shown in Fig. 1.5c.
- *Satin Weave*—The satin weaves are similar to twill weaves, but a number of interlaces of warp and weft are fewer. The 'harness' number is typically 4, 5, and 8. This denotes the total number of yarns crossed and passed under before the fibre repeats the pattern. A crow's foot weave is a type of satin weave that resembles a crow's foot. Satin weaves are very flat due to its long interlace distances and have good

wet finishing and a high degree of drape. The drapeability of a satin weave is considered to be the best. Therefore, satin weaves will adapt to any curvilinear profile during moulding. The flat nature of satin fabric gives low crimps; therefore, its mechanical properties are good. The asymmetry and low stability during handling are the two important aspects of a satin weave. The warp yarn and satin weave interlace over 3, 4, 5, or 8 weft yarns and under 1 weft yarn. The harness of a satin weave is decided by interlacing a warp yarn with the maximum number of weft yarn. It is difficult to handle satin-weave fabric having a harness of 5 or above. Also, care needs to be taken to hide the assembled multiple layers of these fabrics for making composites so that residual stresses are not built into the composite due to the asymmetric effect of weave. The 4-harness satin weave pattern is shown in ◘ Fig. 1.5d.

- *Leno*—A leno weave is primarily developed for locking the open ends of the fabric during weaving. It is also referred to as leno selvedge. It improves the stability in 'open' fabrics that have a low fibre count. Leno is a form of plain weave where twisted warp yarn effectively locks each weft yarn. In this weave, warp yarns are twisted around consecutive weft yarns, forming a spiral pair. Leno weaves are normally used in conjunction with other weave patterns in fabrics in order to avoid their openness so that effective composites are formed. The role of a leno weave becomes very prominent for the handling of satin weaves with a harness 5 and above. The leno weave pattern is shown in ◘ Fig. 1.5e.
- *Mock Leno Weave*—This is also a version of plain weave in which occasional warp yarns deviate from the alternate under overinterlacing of plain weave and rather interlace every two or more weft yarns. The same pattern is repeated in the weft direction, and the resultant fabric has an increased thickness, rougher surface, and additional porosity. The mock leno weave pattern is shown schematically in ◘ Fig. 1.5f.
- *Chopped Strand Mat (CSM)*—CSM is a nonwoven sheet of glass fibre. In CSM, long staple fibres (up to 50 mm) are placed in random orientation in the form of a sheet. The glass strands are adhesively bonded. The glass CSM is available in areal density of 300, 450, 600, and 900 GSM (gram per square metre). The CSM pattern is shown in ◘ Fig. 1.5g. CMS has extremely good adaptability to any shape; therefore, it is used in many applications, such as boat hulls, water tanks, composite furniture, etc. The matrix often used with glass CSM is polyester resin. The products made out of CSM are almost transparent to light.

The comparison of properties with weave styles is given in ◘ Table 1.8.

- *Woven Fabrics vs. Woven Rovings*—Yarn-based fabrics generally give higher strengths per unit weight than roving. These

Table 1.8 Comparison of properties with weave styles [16, 17]

Property	Plain	Basket	Twill	Satin	Leno	Mock leno	CSM
Good stability	2	4	3	4	1	3	3
Good drape	4	3	2	1	5	4	1
Low porosity	3	4	2	1	5	3	5
Smoothness	4	4	3	1	5	4	4
Balance	2	2	2	4	4	2	2
Symmetrical	1	3	3	5	5	2	3
Low crimp	4	4	3	1	2	4	5

Legend: Excellent 1; Very good 2; Average 3; Poor 4; and Very Poor 5

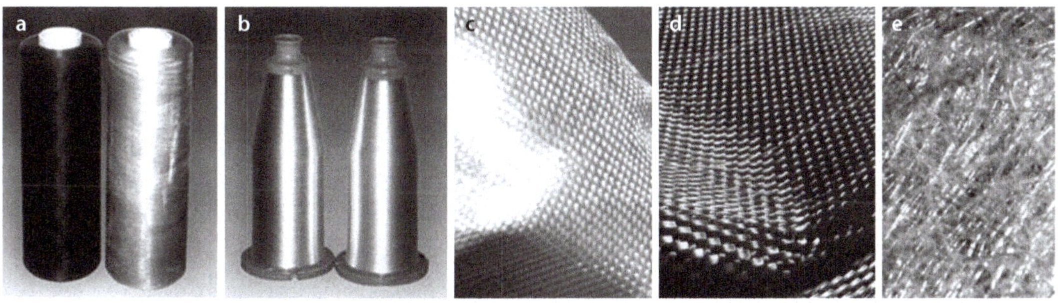

Fig. 1.6 **a** Carbon tow and glass roving, **b** glass filaments, **c** woven glass fabric, **d** woven carbon fabric, and **e** glass CSM [14]. (Image source: Woven Fabrics, ► http://www.netcomposites.com/guide/woven-fabrics/40)

fabrics are generally finer and are available in light-weight ranges. Woven rovings are less expensive to produce and facilitate resin percolation more effectively. Under applied pressure, rovings spread out easily, allowing more resin to percolate and wet them out. Since rovings are available only in heavier texes compared to yarns, fabrics produced out of them are in the weight range of medium to heavy. Therefore, rovings are more suitable for thick and heavier laminates. The commercially available forms of roving, tow, fabrics, and CSM are shown in Fig. 1.6.

1.6 Properties of Matrix Resins

Polymeric resin consists of viscous fluid and solidifies into elastic solids. It responds viscoelastically to applied loads. The role of matrix in fibre-reinforced composites includes the following:

1. Bind reinforcement and fibres in the composite
2. Provide a shape and surface finish to the composite product
3. Transfer stresses between among the fibres
4. Provide a barrier against an adverse environment
5. Protect the surface of fibres from mechanical abrasion

6. Impart rigidity to the composite product
7. Provide a surface for adhesive bonding or mechanical fastening
8. Provide a machineable surface

The matrix could be polymeric, metallic, or ceramic. The most commonly used matrices are polymeric, which are thermosetting and thermoplastic as discussed in previous sections. The properties of popular resin matrices are provided in Table 1.9.

1.7 Properties of Polymeric Composites

The outstanding features of polymeric composites are their high specific strength and modulus and their controlled anisotropy due to their directional properties. The tailorability of their properties makes them very attractive structural materials. Other advantages of polymeric composites are lightweight, high stiffness, good corrosion resistance, impact resistance, fatigue strength, thermal stability, good electrical properties, chemical resistance, and flexibility in design capabilities. Polymeric composites are brittle materials.

Table 1.9 Physical, thermal, and mechanical properties of polymeric matrices [9–12]

Serial no.	Resin matrix	Density (g/cm^3)	Tensile strength (MPa)	Tensile modulus (GPa)	Compressive strength (MPa)	Elongation at break (%)	Service temperature (°C)
	Thermosets						
1.	Epoxy (high- temperature cure)	1.2	80	3	125	5	180
2.	Epoxy (room temperature cure)	1.1	60	2	90	4	100
3.	Polyester	1.2	50	2	130	2	80
4.	Phenolic	1.2	50	5	100	1	125
5.	Polyimide (PMR-PI)	1.4	50	4	120	1	325
6.	Polyurethane matrix resin	1.2	80	2.8	90	5–10	120
	Thermoplastics						
7.	Nylon 6	1.4	65	3	73	52	
8.	High-density polyethylene	0.98	30	1.2	37	55	80
9.	Polypropylene	0.95	35	1.3	42	70	80
10.	Polyetherimide	1.27	100	2.7	125	65	155
11.	Polyether ether ketone (PEEK)	1.32	97	3.6	120	60	156

Table 1.10 Typical physical and mechanical properties of unidirectional (UD) polymeric composites [10–12]

Serial no.	UD composites	Density (g/cm^3)	Tensile strength (MPa)	Tensile modulus (GPa)	Shear strength (GPa)	Poisson's ratio
1.	Carbon/epoxy ($V_f = 60\%$)	1.5	1600	224	0.06	0.28
2.	Boron/epoxy ($V_f = 50\%$)	2.0	1500	148	0.07	0.23
3.	Glass/epoxy ($V_f = 45\%$)	1.9	1300	41	0.07	0.26
4.	Kevlar 49/epoxy ($V_f = 60\%$)	1.4	1380	58	0.03	0.34

Note: V_f is volume fraction of fibres

Their stress–strain curves are linearly elastic to the point of failure by rupture. Polymeric composites may be moulded into any intricate shape, and their properties can be tailored to the specific requirements by careful selection of the fibre, matrix, fibre configuration (short, long, strength, woven, braided, laminated, etc.), and fibre surface treatment.

1.7.1 Physical and Mechanical Behaviour

Polymeric composites have high specific mechanical properties. The specific strength (strength-to-density ratio) and specific modulus (modulus-to-density ratio) of polymeric composites are much higher than those of conventional materials, such as metals. The weight savings due to high specific mechanical properties of polymeric matrix composites make all types of vehicles (such as aircrafts, rockets, automobiles, etc.) higher in fuel efficiency and with a higher capacity to carry a pay load. Typical properties of unidirectional (UD) polymeric composites are given in Table 1.10.

1.7.2 Thermal Behaviour

The thermal performance of polymeric composite largely depends upon the type and chemical nature of the resin matrix and the curing process. Composites present high thermal anisotropy in both along-the-fibre and perpendicular-to-fibre directions. The thermal conductivity ratio between these two directions can vary between 40 and approximately 130. In a unidirectional laminate, the in-plane direction is generally referred to as the *along-the-fibre* direction, and the through thickness direction is often the *perpendicular-to-fibre* direction. In many applications, heat removal is essential in both in-plane and through-thickness directions for effective heat dissipa-

1

Table 1.11 Typical thermal properties of unidirectional (UD) polymeric composites [10–12]

Serial no.	UD composites	Tg (°C)	Thermal conductivity (W/mK)	Flammability (UL-94 rating)	Service temperature (°C)	Degradation T(°C)
1.	Carbon/epoxy (V_f = 60%)	175	165	V1	175	180
2.	Boron/epoxy (V_f = 50%)	175	162	V0	175	180
3.	Glass/epoxy (V_f = 45%)	180	132	V0	180	180
4.	Kevlar 49/epoxy (V_f = 60%)	180	121	V1	180	180

tion. The thermal conduction behaviour of composites having cylindrically orthotropic fibre and an isotropic matrix is fully characterized by the values of its in-plane and through-thickness thermal conductivity [18].

When thermosetting polymers are heated, they will pass through a glass transition temperature, T_g, and eventually thermal decomposition will occur. The properties of thermosetting polymers used in composites are influenced by a high temperature. Therefore, above T_g, the matrix will soften, and the composite experiences a reduction in both modulus and strength. In addition, susceptibility to moisture absorption in the thermosetting matrices is high due to microporosity. As a result, composites are used at temperatures below T_g of the polymer matrix. Thermal decomposition is induced by oxidation. The most common matrix resins are epoxy resins. Epoxies are generally stable up to their T_g, which is 150–180 °C. Oxidation can lead to a change of colour and microcracks. Typical thermal properties of polymeric composites are shown in Table 1.11.

The thermal properties of composites are primarily governed by the matrix used. However, a minor role is played by the reinforcement too. Service temperatures and degradation temperature are only governed by the matrix property in polymeric composites. The temperature profile of thermosetting and thermoplastic composites for their fabrication is pictorially shown in Fig. 1.7. Many resins require a 'post cure' to help reach the highest thermal performance characteristics.

Point to Ponder...

Post curing in thermosetting polymers is akin to annealing in metals. Post curing ensures the complete curing of the thermosetting matrix in polymeric composites and improves the morphology of the matrix. It minimizes the possibility of warpage and dimensional inaccuracy.

A *post cure* is the method of adding temperature for a limited duration of time to the polymeric composite after the resin matrix has already cured through the thermosetting chemical reaction. This helps in completion of residual curing of the matrix resin. A post curing causes optimum cross-linking of a polymeric matrix and organization in polymer chains, increasing structural and thermal properties. The post curing stabilizes the composite product by reducing the internal stresses, reducing its randomness of energy. This random energy is a structural entropy that is reduced after post curing.

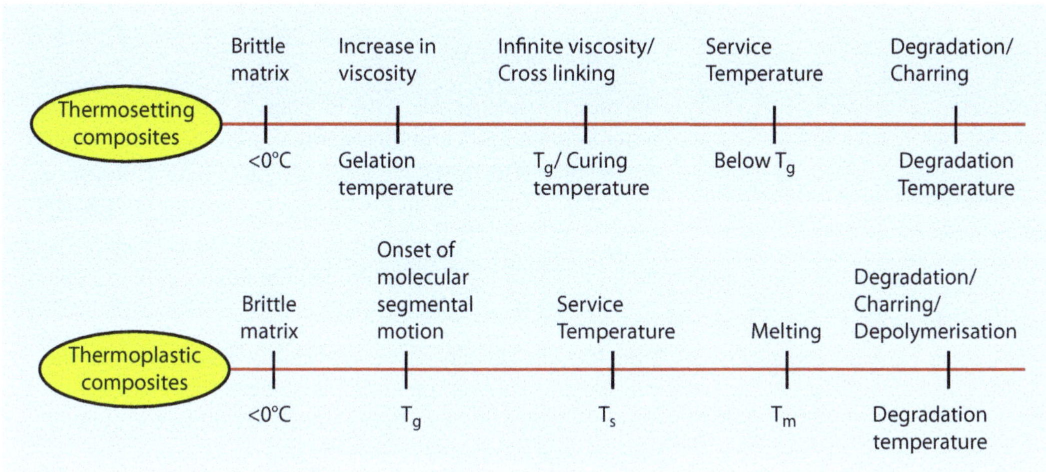

Fig. 1.7 Temperature profile of thermosetting and thermoplastic composites with performance characteristics

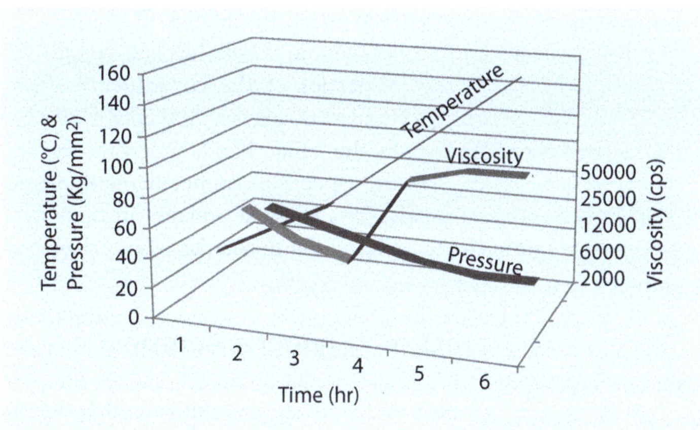

Fig. 1.8 Temperature, pressure, and viscosity relationships in the curing of thermosetting epoxy composite using a compression moulding process

It is to be noted that most composites are susceptible to *ignition*. This results in a flame spreading and the generation of heat and toxic smoke. The reinforcing fibres retard the combustibility of composites using the reinforcement of non-combustible fibres (such as glass fibre). The fibres act as separating agents for the resin, thus restricting the amount of polymer available to fuel the fire.

A typical processing window for epoxy resin-based polymeric composites for the compression moulding process has been depicted in Fig. 1.8. This figure brings out the correlation and interdependency of temperature, pressure, and viscosity with respect to time.

Example 1.5 Prove thermodynamically that high-temperature composites will be more dimensionally stable after post curing than without post curing.

Answer

$$\Delta G = \Delta H + T\Delta S$$

The total free energy, ΔG, is the function of stability of any system, where ΔH, T, and ΔS are enthalpy or total heat content, temperature after post curing, and configurational entropy, respectively. Therefore, for the stability of cured composite, the value of ΔG needs to be minimal.
ΔH is negligible for composites. Hence,

$$\Delta G = T\Delta S$$

Therefore, the stability of the composite will depend upon the entropy, ΔS, while temperature T is variable. The residual energy is stored in the form of entropy. At post curing above T_g, translational movement relaxes the polymeric chains, and residual energy is released so it brings down the configuration entropy.

$$\Delta S \approx 0$$

Therefore,
$\Delta G = 0$, and the post cured composite is dimensionally stable.

1.7.3 Electrical Behaviour

Point to Ponder…
Low dielectric constant and low loss tangent of polymeric composites make it transparent to electromagnetic waves. This property of polymeric composites makes it useful for radar transparent applications in the nosecones of aircrafts and missiles.

Electrical properties of polymeric composites are important to a wide range of industries, such as automotive, aerospace, electrical appliances, marine, packaging, and consumer goods [19]. Electrical tests, in general, are measurements of the resistance, conductivity, or charge storage either on the surface or through the composite material. Typical electrical properties of polymeric composites are given in Table 1.12.

1.7.4 Hygrothermal Behaviour

According to Fick's law, most composites absorb liquid. The *hygrothermal behaviour* of any material is its response to its contact with

Table 1.12 Typical electrical properties of unidirectional (UD) polymeric composites [10–12]

Serial no.	UD composites	Dielectric constant	Loss tangent	Volume resistivity (10^7Ohm.cm)	Surface resistivity (Ohm)
1.	Carbon/epoxy (V_f = 60%)	–	–	V1	175
2.	Boron/epoxy (V_f = 50%)	–	–	V0	175
3.	Glass/epoxy (V_f = 45%)	4.4	0.02	V0	180
4.	Kevlar 49/epoxy (V_f = 60%)	3.5	0.05	V1	180

liquid/moisture under a specific temperature. Composite materials are porous and have breathability. In respect to composite materials, the absorption of only small amounts of liquid may have a significant effect on their performance. The composites get saturated if they are exposed to a liquid for an extended period. The absorbed moisture plasticises and swells the matrix, resulting in a reduction in mechanical properties, glass transition temperature, and service temperature. Hygrothermal ageing deteriorates physico-chemical and thermomechanical properties of the composite over time. Moisture gets absorbed onto the composite via surface defects (such as crazing, cracks, and voids) in the matrix and diffuses into the bulk matrix causing swelling and plasticization. This results in swelling stresses, causing the matrix to crack or the fibre/matrix to debond. Fibre reinforcement is also adversely affected by moisture due to leading effect that causes erosion of the coupling agent or sizing on the fibre surface [20]. For example, water molecules penetrating glass fibre via surface microcracks result in the degradation of the surface treatment on glass fibres, leading to the loss of mechanical properties of polymeric composite.

1.7.5 Chemical and UV Resistance

The chemical resistance of polymeric composites is primarily dependent on the matrix. The role played by the reinforcement is minimal and to the extent that it only allows the ingression of chemicals slowly due to the physical resistance offered by them. In the case of alkali, the glass gets corroded by alkalis; therefore, glass-reinforced composites get corroded by alkalis. Polymeric composites can be designed for excellent corrosion resistance (i.e. resistance to specific hostile environments). The selection of a matrix that is resistant to a specific environment plays an important role to ensure the composite's optimum performance. The selection of reinforcing fibres is the next important thing to make composites resistant to chemicals/UV. The other influencing factors for chemicals/UV resistance of composites are the moulding process, service environment, and expected service life. Once operating and service parameters are established, the appropriate materials can be selected to design the composite [21]. Typical chemical and UV resistance of unidirectional (UD) polymeric composites are provided in ◘ Table 1.13.

All composites are porous and breathable, as mentioned previously. The extent of porosity and breathability varies from materials and processes. Therefore, composites are permeable to water and a number of other liquids to a certain extent. The material selection for composites depends on the severity of the environment in which the composites are likely to be exposed. For a weak acidic environment, chlorinated or long-chain iso polyester would be suitable. Vinyl esters, which have increased chemical resistance, are preferred for aggressive environments of caustics and solvents. Epoxies have some of the best solvent and thermal resistances and are much better than vinyl esters. However, epoxies are comparatively more difficult to process and are more expensive.

1

Table 1.13 Typical chemical and UV resistance of unidirectional (UD) polymeric composites

Serial no.	UD composites	Resistant to chemicals/UV			
		Acids	Bases	Aprotic solvents	UV
1.	Carbon/epoxy ($V_f = 60\%$)	Yes	No	Yes	No
2.	Glass/epoxy ($V_f = 45\%$)	Yes	No	Yes	No
3.	Glass/polyester ($V_f = 45\%$)	Yes	No	Yes	Yes
4.	Kevlar 49/epoxy ($V_f = 45\%$)	Yes	No	Yes	No
5.	Glass/polyimide ($V_f = 60\%$)	Yes	Yes	Yes	Yes
6.	Glass/phenolic ($V_f = 60\%$)	No	Yes	Yes	Yes

The Earth's surface receives UV radiation that has a wavelength range of 290–400 nm, which matches the bond dissociation energy of most of the matrix resins. The UV degradation of polymers is caused by a series of complex processes. UV radiation is absorbed by the polymer, and the UV radiation energy initiates bond dissociation, leading to polymer chain scission and/or cross-linking. In addition, the presence of oxygen can lead to chemical reactions with the formation of some functional groups (such as carbonyl, carboxyl, or peroxide). UV degradation only affects the matrix resins on the surface of a composite; however, the resultant embrittlement of matrix and flaws on the surface layer can provide a point of initiation for fracture of the composite. Only polyesters are inherently UV resistant; otherwise, all other polymers degrade under UV radiation.

The properties of polymeric composites are unique, and these are influenced by the properties of reinforcement and matrix. The parameters of reinforcement and matrix also influence the processability of polymeric composites. In the following chapter, the advantages and applications of polymeric composites are discussed in order to appreciate the potential of polymeric composites in manufacturing industries covering a wide spectrum of applications.

To Be Remembered

1. Composites are the materials/structures that have two or more physically distinct constituent phases. The resultant properties of these materials/structures are superior to the properties of its constituent phases.
2. Anisotropy is the main characteristic of composites.
3. Processability is the capability to transform the given input into a desired output.
4. Drape is the ability of a fabric to conform to a complex 3-D surface.
5. Composites offer a large window of processability; therefore, they can be processed by various processing technologies.
6. Fabric is made out of yarn, and yarn consists of two or three plies that have 'S' and 'Z' types of twists.

7. Strands of glass and carbon fibers are called roving and tow, respectively.
8. Composite materials are porous and have breathability.
9. It is impossible to produce composite free of voids or defects.
10. Most polymers are susceptible to UV degradation.
11. Polyester matrix-based composites are the most UV resistant.
12. Failure in composites initiates from microcrazing or microcracking of the matrix.

Exercises

1. What is processability? What do you understand about the processability of composites?
2. Why is the dimensional stability of a twill or satin weave less than a plain weave?
3. Why are satin weave fabrics having a harness more than 8 not commercialized in composite industries?
4. What are the weaves or pattern of reinforcement to be selected for fabrication of (a) composite boat hull, (b) composite nose cone of aircraft, (c) composite water storage tank, and (d) composite pole/mast? Give reasons for your selections.
5. Calculate the weight percentage of the reinforcement and matrix for (a) glass fibre-reinforced epoxy composite and (b) carbon fibre-reinforced epoxy composites. Densities of glass fibre, carbon fibre, epoxy resin, and polyester resin are 2.54 g/cm^3, 1.8 g/cm^3, 1.2 g/cm^3, and 1.5 g/cm^3, respectively.
6. What should be the critical length of carbon fibre in a short carbon fibre-reinforced bismaleimide composite? The diameter of carbon fibre is 7 μm.
7. What is the failure mechanism in polymeric composites? What are the common defects in composites?
8. Calculate the processability factors for a sisal fibre-reinforced rosin gum natural composite made using a hand layup process. The processable constant of the composite is 0.5. The entropy of this natural composite is 0.1 joules.
9. Calculate the processability constant of an injection moulded E-glass-reinforced nylon composite part. The processability factors, C_f and P_f, are 1.0 each. The value of entropy is 0.2 joules.
10. What is the effect of water on composites? Explain the effect of water on a glass fibre-reinforced polyester composite.
11. What are the frequencies of light that affect composites? What chemical changes occur in composites when exposed to UV radiation?
12. What is the role of anisotropy of a composite in its thermal behaviour? How do you use anisotropy in design to take the advantage of thermal response of the composite?
13. What is the role of a matrix in a composite? How do you select a matrix for a composite?

14. Compare rovings and yarn from the point of view of wettability by resin. Which has more wettability and why?
15. What does post curing do in polymeric composites? What will be the effect on the entropy of the composite after post curing?
16. What are 'S' and 'Z' type of twists in yarn? Explain their response for resin percolation in 'S' and 'Z' twisted yarns during the fabrication of a composite.
17. What are the methods of fabrication of metal matrix composites?
18. What are the specific properties offered by ceramic matrix composites? Compare ceramics with ceramic matrix composites.
19. How does UV radiation affect polymers?

References

1. Hon, D. and Shiraishi, N., eds., *Wood and Cellulose Chemistry*, 2nd ed., Marcel Dekker, New York, 2001.
2. "Notable Progress—the use of plastics," Evening Post, Wellington, New Zealand, Volume CXXVIII, Issue 31, 1939.
3. Shaffer, G.D., "An Archaeomagnetic Study of a Wattle and Daub Building Collapse," *Journal of Field Archaeology*, 20, No. 1 (1993), 59.
4. *Hobart, T., "Car of the future in plastics," The Mercury, 1946.*
5. Hancock, P. and Cuthberston, R.C., "Effect of fibre length and interfacial bond in glass fibre-epoxy resin composites," *Journal of Material Science*, 5 (1970), p. 762.
6. Abdul Kalam, A.P.J. with Tiwari, A., *Wings of Fire, An Autobiography*, University Press, India, 1999.
7. Composite boat hull, *Composites World*, 19 January 2009.
8. Unsyayed Mast, http://en.wikipedia.org/wiki/Unstayed_mast.
9. Varma, I.K. and Gupta, V.B., *The Science and Technology of Fibrous Composites*, Asian Books Pvt Ltd., New Delhi, 2014.
10. Agarwal, B.D. and Broutman, L.J., *Analysis and Performance of Fiber Composites*, John Wiley & Sons, USA, 1980
11. Bunsell, A.R., *Fiber Science and Technology*, CRC Press, USA.
12. Mohanty, A.K., Misra, M., and Drzal, L.T., Ed., *Natural Fibers, Biopolymers, and Biocomposites*, CRC Press, Taylor and Francis, 2005.
13. Pludeman, E. and Collins, N., *Adhesion Science and Technology*, Vol. 9a, Plenum Press, New York, 1975.
14. Woven Fabrics, http://www.netcomposites.com/guide/woven-fabrics/40.
15. Woodhams, R.T. and Xanthos, M., *Handbook of Fillers and Reinforcements of Plastics*, Ed. Katz, H.S. and Milevishi, J.V., Van-Nostrand, Princeton, NewJersey, 1978.
16. Alif, N., Carlsson, Leif, A., and Boogh, L., "The effect of weave pattern and crack propagation direction on mode I delamination resistance of woven glass and carbon composites," *Composites Part B*, 29B (1998), p. 603.
17. Suppakul, P. and Bandyopadhyay, S., "The effect of weave pattern on the mode-I interlaminar fracture energy of E-glass/vinyl ester composites," *Composites Science and Technology*, 62 (2002), p. 709.
18. Hasselman D.P.H., Donaladson H.K.Y. and J.R.J. Thomas, "Effective thermal conductivity of uniaxial composite with cylindrically orthotropic carbon fibers and interfacial thermal barrier," *Journal of Composite Materials*, 27 (1993), p. 637.
19. Smallman, R.E. and Bishop, R.J., *Modern Physical Metallurgy and Materials Engineering*, 6th ed., Oxford, Butterworth-Heinemann, 1999.
20. http://www.easycomposites.co.uk/products/chopped-strand-mat/450gemulsion-bound-csm-per-metre.aspx.
21. Smithers, Rapra, http://www.rapra.net/composites/material-selection/some-design-considerations.asp.

2

Advantages and Applications of Polymeric Composites

A. K. Ghosh, M. Dwivedi, *Processability of Polymeric Composites*,
https://doi.org/10.1007/978-81-322-3933-8_2

Composites have engineering advantages over conventional materials, such as metal, wood, and leather. Polymeric composites have good tensile strength, flexural strength, compressive strength, impact strength, Young's modulus, and rigidity coupled with high dimensional stability. Depleting natural resources and related environmental concerns—specifically in the use of wood—have forced scientists and engineers to explore more applications for advanced composites. It can be said that advanced composites have come to a level where they can replace conventional materials in most engineering applications [1, 2]. Composites have proven themselves to be the best substitute for conventional materials. The advantages of polymeric composites and their applications are discussed in detail in the following sections of this chapter.

2.1 Advantages of Polymeric Composites

The major advantages offered by polymeric composites are in their performance, cost, and production rate. However, the main advantages polymeric composites are discussed here.

1. *High Specific Strength and Modulus*—The specific strength and modulus of any material are its strength-to-weight ratio and modulus-to-weight ratio, respectively. This is generally calculated by dividing strength or modulus by the gravity of the material. The specific strengths and modulus of composites are much higher than those of metals. Conventional materials, such as steel, are very strong and heavy. Composite materials can be designed to be both strong and light in weight. The property of high specific strength makes composite preferred materials to build aircrafts that need a very high-strength material at the lowest possible weight. A composite can be made to resist bending in one direction by reinforcing it more at that particular location [3–5]. Whereas in metals, extra thickness or a new design feature to increase moment of inertia would be required to improve the bending strength, which will add weight. Composites can be made strong without being heavy. Composites have the highest strength-to-weight ratios in all structural materials available to date [6–9]. Properties of composites and conventional materials for their specific tensile strength and tensile modulus are shown in ◘ Fig. 2.1

   Composites have high strengths, and they can be designed to be far stronger than aluminium or steel. In view of the fact that composites are anisotropic in nature, they have maximum strength in the direction of the fibre [10–12]. Composites can be engineered and designed to be strong in a specific direction. Composites have high stiffness, strength, and toughness and are often comparable with structural metal alloys (such as steel, aluminium, etc.), as discussed in ► Chap. 1. Low densities of composites are important for energy-efficient systems and platforms, such as automobiles and aircraft [12, 13]. For example, a bus with a body made out of composites is operating on London roads, as shown in ◘ Fig. 2.2.

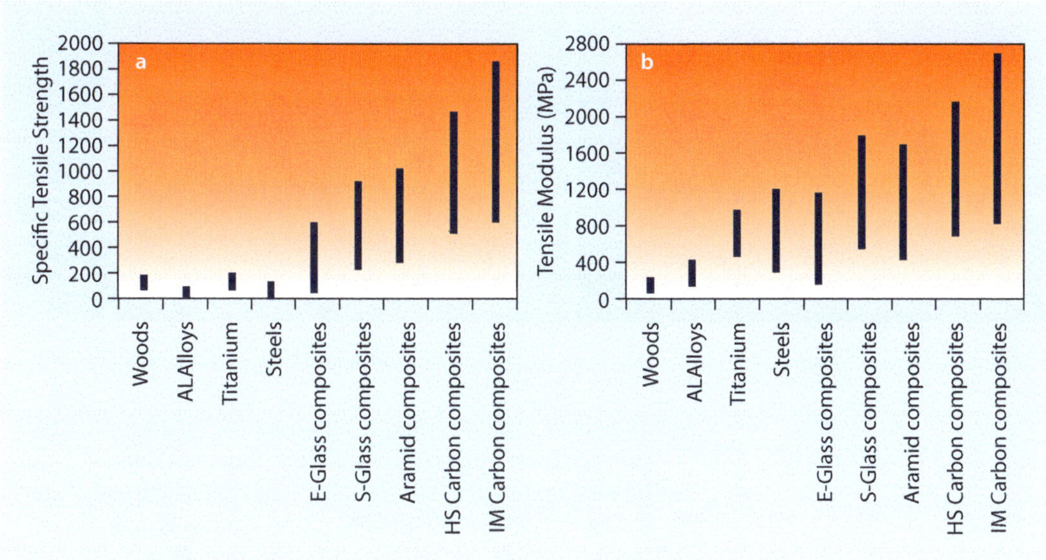

Fig. 2.1 Properties of composites and conventional materials: **a** specific tensile strength and **b** tensile modulus

Fig. 2.2 Bus having a body made out of composites [14]

Less weight in the composite structure for automobiles and aircrafts means better fuel efficiency. Fuselage and airframes of modern airplanes (such as the Boeing 787 Dreamliner, B-52 bomber, etc.) are built using carbon fibre-reinforced epoxy composites. In automotive industries, natural fibre-reinforced composites (NFRC) are also becoming popular. Mercedes Benz S class is using 27 parts made out of NFRC. Flax fibre-reinforced PP instrumental panel is shown in Fig. 2.3a. Even cotton fibre-reinforced composites also are being used in automotive industries. In the Toyota Lexus and Camry Mark X, kenaf fibre-reinforced PP is used in the inner board of door panels, as shown in Fig. 2.3b. This has reduced the weight of door panels by 20%.

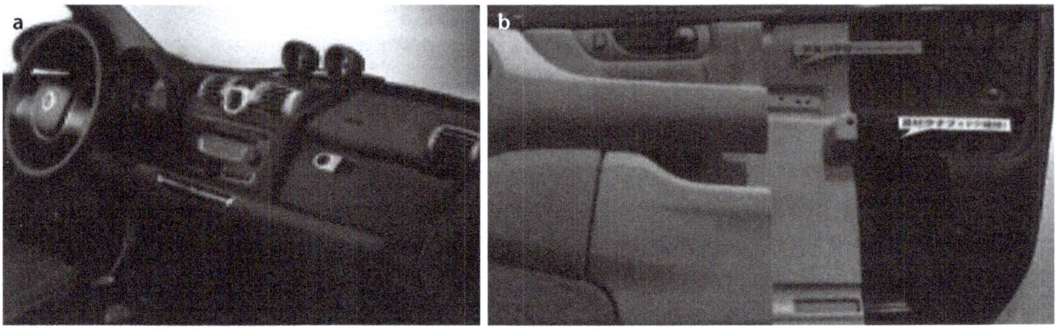

Fig. 2.3 a Flax fibre-reinforced PP instrumental panel and b kenaf fibre-reinforced PP inner board of door panel [15]

2. *Tailorability of Properties*—Composites are anisotropic (i.e. have different properties in different directions), and this can be used to design a more efficient structure. Composites are the materials in which desired properties can be tailored as per the requirements of the load/deflection. Composites provide the possibility to reinforce more fibres and/or variety of fibres to achieve the desired strength or stiffness within the specified volume or weight of the structure. These basic properties of composites keep them above other materials. These are the reasons composites are not referred to as materials by engineers. Rather, they are considered structures. Every composite structure has its own characteristic and uniqueness in terms of desired properties due to their anisotropy [16, 17]. This uniqueness is not possible with isotropic materials such as metals. In many structures, the stresses are also different in different directions. For example, in closed-end pressure vessels (such as a rocket motor case), the circumferential stresses are twice the axial stresses. Using composites, such a vessel can be made twice as strong in the circumferential direction as in the axial.
3. *Improved Heat Deflection Temperature*—The heat deflection temperature is a measure of the ability of thermoplastic composites to bear the load at elevated temperatures. The deflection temperature is also known as the deflection temperature under load (DTUL), heat deflection temperature, or heat distortion temperature (HDT). The reinforcing fibres restrict the distortion of composites under heat and load. However, residual stresses in the composite structures cause distortion, and such distortions are more prominent in thin sections of polymeric composites. The deflection temperature is a useful measure of the relative service temperature for load bearing composites. The comparison of HDT between composites and their matrix resin is given in Table 2.1.

   However, the deflection temperature test is a short-term test and should not be used alone for product design. Other factors such as the time of exposure to elevated temperature, the rate of temperature increase, and the part geometry all affect the performance.

Table 2.1 Comparison of HDT between composites and their matrix resin [18]

Serial no.	Material	Deflection temperature at 0.46 MPa (°C)	Deflection temperature at 1.8 MPa (°C)	Melting point (°C)
1.	Abs	98	88	–
2.	ABS + 30% glass fibre	150	145	–
5.	Nylon 6	160	60	220
6.	Nylon 6 + 30% glass fibre	220	200	220
7.	Polyethylene terephthalate (PET)	70	65	250
8.	PET +30% glass fibre	250	230	250
9.	Polypropylene	100	70	160
10.	Polypropylene +30% glass fibre	170	160	170

Table 2.2 CTE of composites [19]

Serial no.	Material	CTE (10^{-6}/°C)
1.	Carbon/epoxy	−0.002
2.	Glass/epoxy	−0.414
3.	Graphite/epoxy	−1.051

4. *Low Coefficient of Thermal Expansion (CTE)*—Composites have a low coefficient of thermal expansion (CTE); therefore, they don't cause any problems with dimensional stability on the application of temperature at or below their T_g. Most of the reinforcing fibres have a low (but positive) linear coefficient of thermal expansion. Carbon fibre has *negative* linear CTE, but this shrinkage is compensated in composites due to their positive linear coefficient of thermal expansion of polymeric matrix. Mostly, composites have negative values of CTE, as shown in Table 2.2.
5. *Increased Mechanical Damping*—Damping in structures is commonly provided by viscoelastic materials. The damping characteristic of composites is substantial due to a viscoelastic matrix. The interfaces and viscoelasticity of the matrix in composites increase the damping capacity of composites with negligible decrease—if any—of the storage modulus. Significant damping capacity along with high strength and stiffness is desired in composite structures. Composites generally offer relatively high levels of damping. Furthermore, the damping often can be tailored to desired levels by using suitable raw materials and processing.
6. *Corrosion Resistance*—Composites are made of polymeric matrices and reinforcing inorganic or organic fibres. The matrix is the outermost surface of the composites, and it is corrosion resistant. Therefore, they resist damage from the

Fig. 2.4 Composite windmill blade [16]

Fig. 2.5 Lightweight composite boat [17]

weather and from harsh chemicals that can corrode metals. Composites are good choices for storing and handling chemicals. Outdoors, they stand up to severe weather and climatic thermal shocks. Windmill blades and yachts are the testimony of the corrosion resistance of composites. The longest ever windmill blade has been developed by M/s Vestas Wind Systems, Denmark. It is made out of glass fibre-reinforced composites, as shown in Fig. 2.4. The blade has a length of 80 metres, and the swept area of the rotors is 21,124 m^2. It has been designed for 25 years of lifetime.

A lightweight composite boat weighing 40 kg is shown in Fig. 2.5. Its composite hull provides corrosion resistance from seawater.

Point to Ponder…

Soft armour panel (SAP) used in bullet proof jackets is made out of loosely stacked unidirectional sheets of UHMWPE fibres. SAP is capable of stopping bullet fired from a pistol.

7. *High-Impact Strength*—Composites can be made to absorb impacts due to the viscoelastic nature of the matrix and high tenacity of the reinforcing fibres. Because of this property, composites are used in high-impact-resistant applications in aircrafts, buildings, and vehicles to protect them from shock and impact damage. The pliability of the interphase of the composites makes it more resistant to impact. Very strong bonds between the fibres and the matrix break at small strains. Therefore, this relatively weaker bond between the matrix and the fibres makes the entire structure impact-resistant due to its capability to yield under stress or strain. The high-impact resistance also makes the composite shock resistant. A common example of impact resistance is the bullet-resistant

Fig. 2.6 Bullet-resistant composite helmets [18]

Fig. 2.7 CFRP bogie frame [20, 21]

composite helmets shown in Fig. 2.6. These helmets are made out of Kevlar (Kevlar 29 or Kevlar XP, which are high-impact-resistant fibres) fibre-reinforced epoxy composites. Polymeric composites are used in blast-resistant structures also.

8. *High Fatigue Resistance*—Many structures experience fatigue loading where the internal stresses vary with time. For example, Kawasaki Heavy Industries in Japan has developed a next-generation rolling stock bogie frame, as shown in Fig. 2.7. It uses CFRP for its main structure framework, which also has CFRP leaf springs. The suspension of the bogie eliminates the need for coil springs. This has reduced the weight of the bogie by approximately 40% compared to that of a steel frame. The material having a higher strength and lower modulus of elasticity is the most suitable material for the leaf springs. In each spring, potential energy is stored as strain energy, and then it is released slowly.

   The higher energy storage capability of a leaf spring ensures a more efficient suspension system. The strain energy storage capability of leaf springs is provided in Table 2.3.

   It can be observed from Table 2.3 that the composites make better leaf springs. Another parameter of importance in leaf springs is fatigue life. In *fatigue*, the stresses vary sinusoidally from tension to compression as the axle turns, and fatigue

2

Table 2.3 Strain energy stored by material (kJ/kg)

Serial no.	Materials	Strain energy stored by material (kJ/kg)
1.	Steel (EN47)	0.3285
2.	Carbon/epoxy	2.45
3.	Carbon/polyurethane	4.12
4.	E-glass/epoxy	4.5814

Fig. 2.8 Composite leaf spring for new Volvo SUV XC 90 [22]

stresses are created. These fatigue stresses can eventually lead to failure—even when the maximum stress is much less than the static failure strength of the material. Composites in general and carbon fibre-reinforced composites in particular have excellent fatigue resistance in comparison to metal alloys. Leaf springs made out of carbon fibre-reinforced polyurethane matrix composites are being used in the Volvo SUV XC 90, as shown in Fig. 2.8.

Composites often show accumulated fatigue damage. As a result, this damage can be detected in advance, and the part can be replaced before a catastrophic failure occurs. Materials also have damping characteristics, and some fraction of the mechanical strain energy is deposited in materials by a fatigue loading cycle. This energy is dissipated as heat, especially in the case of metals. This phenomenon can be advantageous, for instance, in controlling mechanically induced vibrations [23–25]. While composites generally offer relatively high levels of damping, the dissipation of heat is limited due to their dielectric nature.

Point to Ponder...
Tribology is the study of interacting surfaces in relative motion caused by friction, lubrication, and wear.

9. *High Tribological Resistance*—Composites can be excellent materials in applications involving sliding friction. Their tribological (wear) properties approach those of lubricated steel. CF-reinforced rigid polymeric composites (such as carbon–carbon composites) are best suited for lubrication purposes. A stack of CFRP leaf springs in the shock absorber of any vehicle is also an example of self-lubrication among leaf springs during their loading and unloading [26, 27].
10. *Design Flexibility*—Composites can be moulded into complicated shapes more easily than conventional materials. This gives designers the freedom to create almost any shape or

Fig. 2.9 a Wing assembly and **b** rudder assembly [28] racquets, etc. are made from carbon or glass fibre-reinforced composites because these materials easily can be moulded into complex shapes. The surface of composites also can be moulded into any texture—from coarse to smooth

form. A wing and rudder assembly made out of composites is shown in Fig. 2.9. Sports boats, bicycles, tennis

11. *Dimensional Stability*—Composites retain their shape and size in any environment, such as cold, hot, arid, and humid. Natural fibre-reinforced composites, on the other hand, may swell and shrink as the humidity changes. Composites are the choicest of materials that have high dimensional accuracy. The excellent dimensional accuracy of composites is attributed to their low linear thermal expansion, nonconducting characteristics, and high rigidity.
12. *Nonconductive Characteristic*—Composites are nonconductive due to the very nature of their constituent phases that include nonconductive polymers and fibres. Only carbon or graphite fibres conduct electricity. The nonconductivity of composites makes them suitable for use as electric poles and for electronic circuit boards. If electrical conductivity is needed, composites may be designed to be electrically conductive by reinforcing conductive materials and metal powders, such as silver powder. The housing of conductive composites is used in electromagnetic inductance (EMI) and electromagnetic conductance (EMC) for the shielding of electronic circuits.
13. *Nonmagnetic Characteristics*—Composites are nonmagnetic as they do not contain any magnetic material or metallic substance. This is a unique property of composites and allows them to be used in both electronics and electrical designs. The nonmagnetic character of composites makes them very good materials for constructing the hulls of stealth frigates and boats.

Point to Ponder…

The magnetic signature of ferromagnetic material in the sea can be detected very easily because there is no magnetic material present in the sea except in boats, ships, and submarines.

14. *Radar Transparency*—Composites may be designed with a low dielectric constant and a low loss tangent so that they will cause a minimum attenuation of EM radiation to pass through them. This property makes composites ideal materials for application in radomes and nose cones. The composite will protect the antenna from damaging operational and environmental effects, while at the same time allow an EM signal to pass through it. Radar-transparent composites are made from glass or Kevlar fibre-reinforced polymers, such as epoxy resin, low loss polyester resin, phenolic resin, or polyimide resin. Carbon fibres are never used as reinforcement for radar-transparent applications due to their conducting and semiconducting behaviour. Radio-frequency identification devices (RFID) used for high frequency (HF), very high frequency (VHF), and ultra-high frequency (UHF) that are embedded in composites are used in various applications.
15. *Low Thermal Conductivity*—Composites are good insulators due to their constituent phases, which do not easily conduct heat. Composites exhibit a low thermal conductivity because of their relatively low density, weak chemical bonding, amorphous structure, and high anharmonicity in their molecular vibrations. The main mechanism of heat conduction in most polymers is through phonons. The interaction among phonons and their interaction with subatomic particles cause a transfer of heat energy. The phonon scattering is caused by anharmonicities due to lattice imperfections, voids, and impurities.
16. *Durability*—Structures made of composites are durable and have a long lifespan. They need little maintenance. Polymeric composites may be designed for required performance. By using a hybrid reinforcement and the choicest of matrices, polymeric composites have the possibility of ensuring that a product lasts for its designed lifespan.
17. *Fire Retardancy*—The fire resistance of composites is dependent on the fire resistance of the matrix. Common matrices are not inherently fire-resistant. Therefore, fire retardants are added to make it fire-resistant. The matrices containing ether group (-O-) are inherently fire-resistant, and these matrices are PEEK, PEI, etc. Among reinforcements, glass fibre is fire-resistant, whereas carbon and graphite fibres are not.
18. *Ablation Property*—Composites may be designed for ablation applications. Common reinforcements for ablative composites are glass fibres (preferably with high silica content) and carbon fibres (preferably rayon based). The preferred matrices for ablative composites are phenolic resin, epoxy resin, and polyimide resin. The mechanism of energy exchanges in the ablation process is shown in Fig. 2.10.

    Among all matrices, phenolic resin is used the most because of its good ablative properties. Ablative composites are used in re-entry aerospace vehicles, as liners for rocket motors, in hypersonic vehicles, etc.

Point to Ponder...
***A**blative composites made possible the manufacturing and flying of space shuttles, rockets, and intercontinental ballistic missiles (ICBM).*

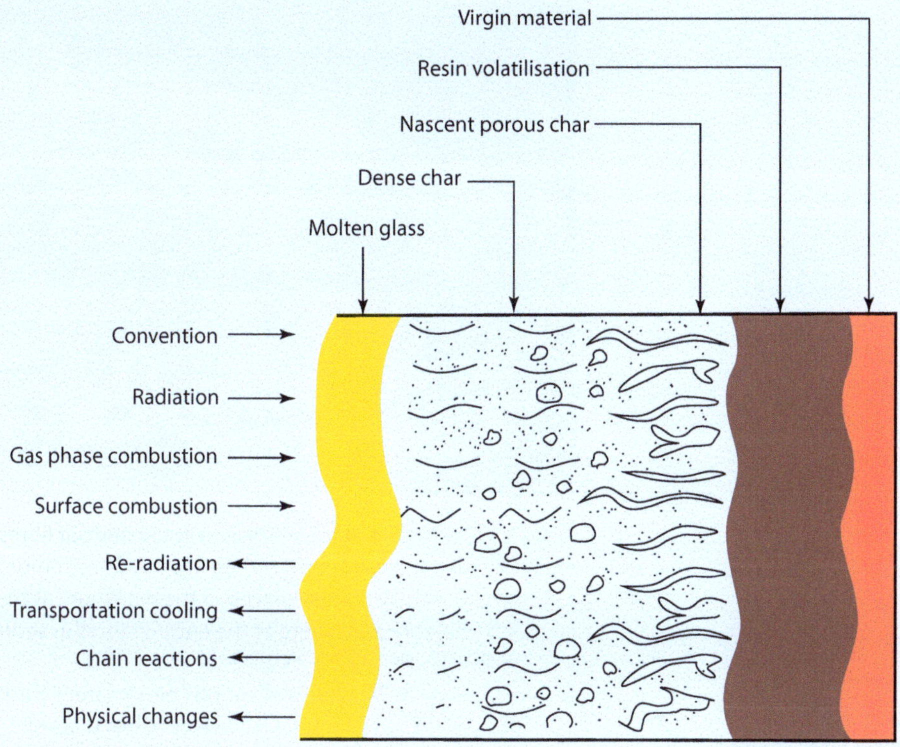

Fig. 2.10 Energy exchanges in the ablation process

19. *In Situ Mouldability*—Composites provide the possibility of in situ moulding of any part, such as a metallic insert, bulkhead, lug, nut, bolt, washer, etc. The in situ moulding provides a strong bond between the metallic insert and the composite without requiring any secondary process for the bonding of the metallic inserts. Secondary operations not only increase production time (due to operations such as machining, drilling, etc.) to fit the metallic inserts, but they also weaken the composite structure. The in situ moulded metallic bulkhead in the composite fuselage of an aircraft is shown in Fig. 2.11
20. *Amenable to Pigmentation and Colouring*—Most of the composites may be pigmented by mixing a pigment into the matrix. Also, all composites may be painted many desired colours. Sporting goods, leisure equipment, etc. that are made out of carbon fibre-reinforced composites are usually painted to improve their aesthetic look rather than keep their inherent black colour.
21. *Heat Resistance*—Polymeric composites are heat-resistant up to the service temperature of their matrices. Heat resistance is different from HDT. Heat resistance implies that the structural integrity of a composite remains intact upon continuous exposure of high temperature without delamination and loss of mechanical properties. Composites made out of high-temperature matrices (such as PMR-PI, PEEK, PEI, PAI, etc.) have high temperature resistance. PMR-PI composites may

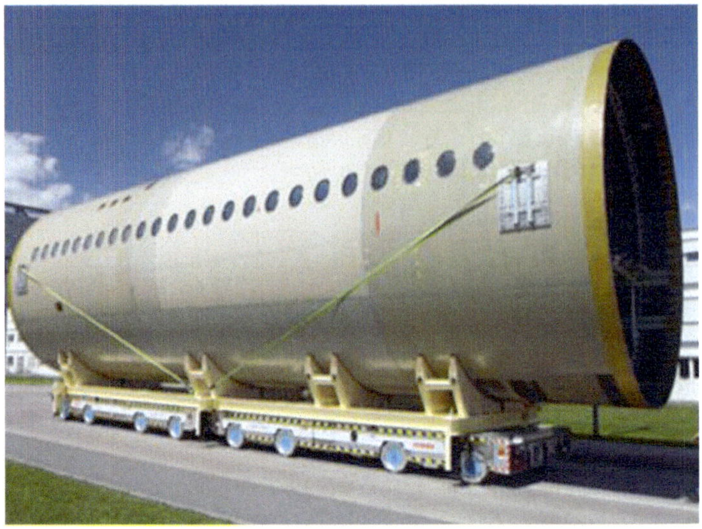

Fig. 2.11 In situ moulded metallic bulkhead in the composite fuselage of an aircraft [29]

provide service temperatures in excess of 300 °C. It is used in supersonic aerospace vehicles.

22. *Cost-Effectiveness for Small Quantities*—Fabrication of high-quality products in small quantities or of one odd item is more cost-effective in composites rather than making the same product out of conventional materials. This is due to the low cost of tooling and processing involved in the fabrication of composites. Toolings made out of composites, plaster of Paris, wood, etc. are cost-effective for limited production of composite products.
23. *Low Processing Cost*—Composites have lower processing costs and lower wastage compared to conventional materials. The infrastructure required to process composites is also minimal due to fewer numbers of processing steps. The durability of the tooling for high production volumes of composites is quite good, and the cost of tooling is amortized in a few initial production lots.
24. *Low Material Cost*—The consumption of composite materials is low due to their high specific strength and modulus. This makes overall costs of composites quite low. Also, composites can be precisely moulded; therefore, there is no or little wastage. Composites do not rust as do many ferrous alloys, and resistance to this common form of environmental degradation may offer a better life cycle cost—even if the original structure is initially more costly.
25. *Reduction in Inventory*—A single piece made of composite materials can replace an entire assembly of metal parts. Reducing the number of parts in a product or a structure saves time and cuts down on the maintenance and cost needed over the lifetime of the composite product. Assembly from a number of subassemblies adds to the process and complexity of the design. Composites offer a lot of flexibility in processing

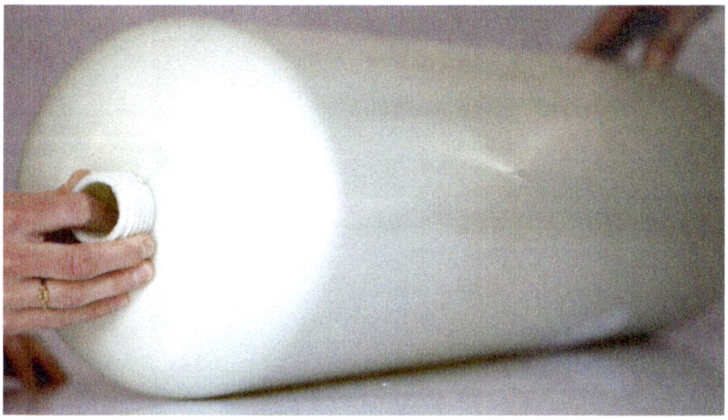

Fig. 2.12 Tape wound composite CNG storage tank [30]

and property control, and this often leads to possibilities for reduction in subassemblies and simple manufacturing. For example, a multistep process of manufacturing a metallic high-pressure vessel may be overcome by a single process of filament winding. A thermoplastic, filament wound composite compressed natural gas (CNG) tank (type IV pressure vessel) has been made using tape winding over a thermoplastic (nylon 6) liner. The liner is blow moulded and 2 m in length, as shown in Fig. 2.12. The tape winding is carried out using glass- or carbon-reinforced epoxy prepreg tape.

26. *Improved Productivity*—The assembly costs of composites may be reduced by combining several previously assembled parts into a single component. Injection moulding and pultrusion processes have very high production rates. Even so, the compression moulding process with multi-daylight moulds also may provide a good rate of production. In addition, composites do not usually require additional machining, thereby reducing work-in-process and time-to-market concerns.

2.2 Disadvantages of Composites

Composites are not perfect for all applications, and the designer needs to be aware of their drawbacks as well. The disadvantages of composites are enumerated here.

1. It is ironic (but true) that variation in composite products from process to process and batch to batch does occur. This sometimes creates an apprehension of the use of composites in the minds of designers.
2. Anisotropy is advantageous because it provides great design flexibility, but the downside is that they also complicate the design. The well-known software and tools of stress analysis used in isotropic linear elastic design need to be customized or modified for anisotropy.
3. Composites are heterogeneous due to anisotropy; therefore, properties in composites vary from point to point in the material. The relationship between stress and strain (force and

Point to Ponder...
Composites are known to fail catastrophically due to brittle fracture. The catastrophic failure occurs from microcrazing of the surface resin, then by the debonding of fibres, and followed by delamination of laminae.

2

Fig. 2.13 Torn off tail fin of an aircraft [31]

deformation) is much more complicated in anisotropic materials, often causing composites to fail catastrophically. The broken composite tail fin of an aircraft is shown in Fig. 2.13.

4. All applications are not weight-critical. If weight-adjusted properties are not relevant, steel and other traditional materials may work fine at a lower cost.
5. Most composites require a hot curing process. Therefore, specialized tooling and processing equipment are necessary.
6. Riveting and the placement of screws require special attention so that unnecessary stresses are not created in the composite.
7. Repairs in composites are difficult and pose new problems. Composites must be thoroughly cleaned of all contamination before repair. The most common repair technique for noncritical composite parts and assemblies is patching by contact layup. The stepped contact layup is another important technique for the repair of composites. In this technique, the cracked portion is cut into steps parallel to the cut. The smallest and first step is made near the crack. The width of this first step is only 10 mm on both sides of the crack. The next step is made symmetrical to first step with 20 mm on both the sides of the crack, the third is 30 mm, and so on. Each step is made by cutting one or two reinforced layers. Therefore, the number of steps is governed by the repair plan. Every step created is then thoroughly cleaned by acetone and dried. The gaps created by the steps are filled with a resin pre-impregnated reinforcing fabric. Two layers of resin-impregnated reinforcement are placed on both faces of the crack. Finally, repaired crack will be cured under pressure and/or temperature using a jig or fixture. The schematic of this repair plan is shown in Fig. 2.14.
8. Composite repair requires a plan that is unique to every repair and requires special tooling and materials. The repair of some of the composite assemblies and parts can be carried out using the hot bonder process shown in Fig. 2.15.

Point to Ponder…

Repairing composites is a highly skilled job. It needs a thorough analysis of the defected part and a repair plan.

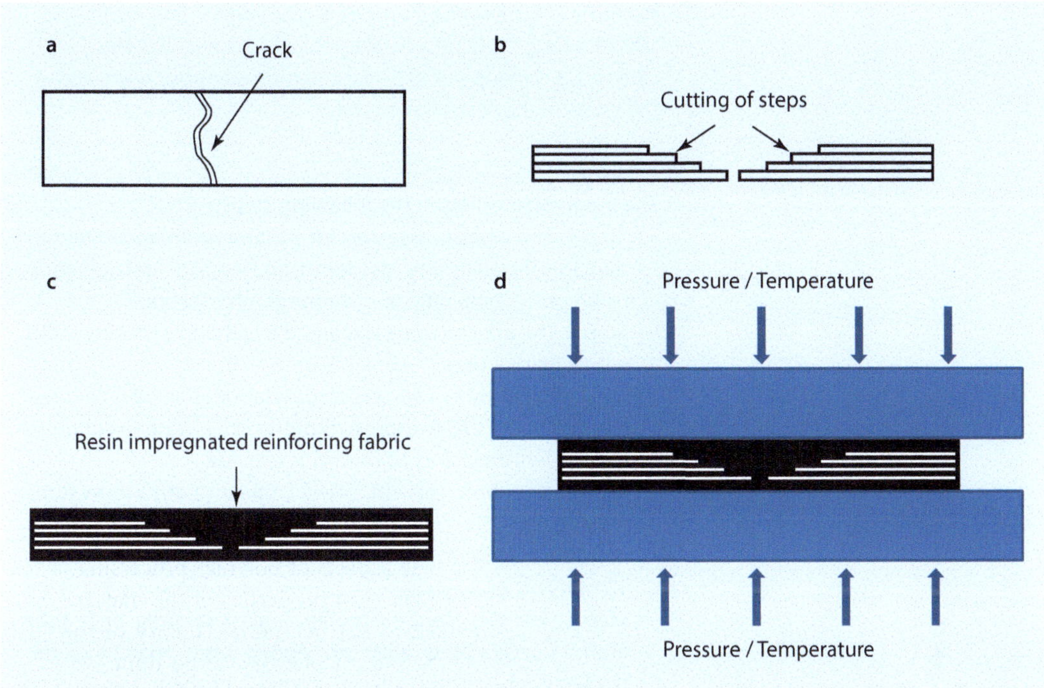

Fig. 2.14 **a** Cracked composite, **b** cutting of steps at cracked surface, **c** placement of resin-impregnated reinforcing fabric, and **d** curing of composite layup

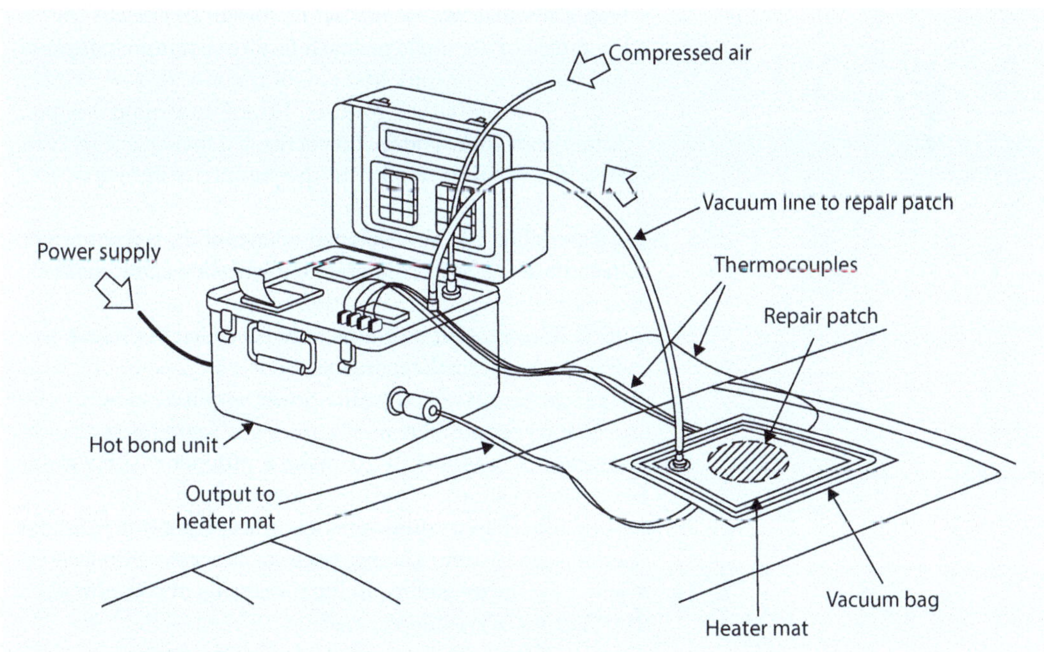

Fig. 2.15 Typical hot bonder portable unit [32]

Example 2.1 How will you repair 5-mm-wide, 6-mm-deep, and 100-mm-long crack on a fuselage of a cargo aircraft? The composite is made out of S-glass-reinforced epoxy resin, and thickness of the section is 15 mm.

Answer

The repair plan will have the following steps:

- Step 1. The crack is assessed for its dimension and extent of damage by analyzing its depth and area of delamination using pulse-echo ultrasound sonographic testing.
- Step 2. The affected area is marked by adding a 25 mm width with a marker pen.
- Step 3. The number of layers of glass fabric affected is counted (say *n* number). At the location of the crack, a clean cut up to $n + 2$ layers deep and 10 mm wide is made.
- Step 4. Now from this cut, every lamina is cut to form steps leading up to the top of composite. The width of each step is measured as Width of every step = (total marked width – 10 mm)/no. of laminae
- Step 5. For this stepped profile, S-glass prepreg or S-glass fabric impregnated with an epoxy resin matching the thickness of lamina is placed stepwise in the repair area so that the prepared profile is completely packed.
- Step 6. One extra layer is placed on the top in order to exert positive pressure during curing.
- Step 7. The entire assembly is prepared for vacuum-assisted curing.
- Step 8. After curing, the top layer is filed or emeried to match the profile of the fuselage, and a final coat of resin is applied and cured.
- Step 9. The assembly is checked for any flaw using another pulse-echo ultrasound flaw detector.
- Step 10. If no flaw is found, the assembly is ready for use.

9. The experience and intuition gained over the years about the behaviour of metallic materials do not apply to composite materials.
10. Carbon fibre-reinforced composites have negative linear coefficient of thermal expansion.
11. The raw materials of composites must be dried before processing because all resin matrices and some fibres absorb moisture.
12. The curing of some composites poses health hazards.
13. Carbon fibre-based composites are inherently black in colour. Therefore, they cannot be pigmented.
14. Monitoring the overall health of composites and prediction of their lifespan are a complex task. In November 1999, the America's Cup boat 'Young America' broke in two due to the debonding face/core in the sandwich structure.
15. Even after several years of emphasis on composites as the 'material of the future', economies of scale are still not well

developed. As a result, high-performance composites are expensive. It is essential to develop newer technologies to bring down the cost of raw materials and processing. Of course, designers need to look at various advantages of composites vis-à-vis the cost [33]. During the energy-crisis period of the 1970s, automobile manufacturers were so anxious to reduce vehicle weight that they were willing to pay a premium for composites for their advantage having lower weight. However, as soon as society's worry about energy efficiency diminished, the industry gradually returned to a strict lowest-cost approach in selecting materials. Hence, the market for composites in automobiles returned to a more modest rate of growth. Although composites have been used extensively in structural applications, the long-term durability of these materials is less than that of steel or other traditional structural materials.

2.3 Applications of Polymeric Composites

The applications of polymeric composites are diverse and have a deep outreach into society. They find applications in all industrial sectors ranging from space shuttles and fighter aircraft to furniture and bathtubs. In an ever-growing worldwide market of over 50,000 different types of components, it is nearly impossible to provide a complete list of the applications of composites. However, some of the applications classified according to major market segments are mentioned here.

2.3.1 Agriculture

Containers and enclosures, equipment components, feed troughs, fencing, partitions, flooring, staging, silos, and tanks used for agriculture may be made of composites. Today greenhouses are constructed using glass CSM-reinforced polyester resins. The inherent UV resistance of polyester provides UV stability and durability to the greenhouse. The greenhouses are made of CSM or polyester composites. A typical greenhouse in the Netherlands is shown in ◻ Fig. 2.16.

2.3.2 Aviation

The use of fibre-reinforced composites has become an increasingly attractive alternative to conventional metals for many aircraft components, mainly due to their increased strength, durability, corrosion resistance, resistance to fatigue, and damage tolerance. Boeing is developing a composite wing with a length of 233 ft. for their Boeing 777X aircraft. It is going to be the largest wing made so far in composites' history. Composites are used in the construction of

Fig. 2.16 Greenhouse [34]

Fig. 2.17 Boeing 787 Dreamliner aircraft [35]

the airframe of the Boeing 787 Dreamliner aircraft shown in Fig. 2.17. The use of composites reduces the weight of aircraft and improves their fuel efficiency.

Composites with advanced technologies (such as 3-D woven fabric-reinforced composites, hybrid composites, UV-resistant skins, electrically conductive composites, smart and intelligent composites, etc.) will facilitate faster construction of future aircraft. Aeronautic structures (such as fuselage, control surfaces, light aircraft construction, internal fittings, partitions and floors, window masks, galley units and trolleys, structural members, satellite components, aerials and associated enclosures, ground support equipment components, etc.) may be readily replaced by composites.

The Robo-Raven is a micro unmanned aerial vehicle (UAV) shown in Fig. 2.18. It is made out of composites, weighs 9.7 g, and has a wing span of 34.3 cm. This UAV can carry a payload of approximately 6 g. It is made out of carbon fibre, 3-D printed lightweight thermal-resistant plastic, Mylar® (a type of polyester) film, and foam. Robo-Raven glides, soars, and flaps like a real bird. It has the remarkable ability to hide in plain sight due to its small size.

Fig. 2.18 The Robo-Raven micro air vehicle (MAV) [36]

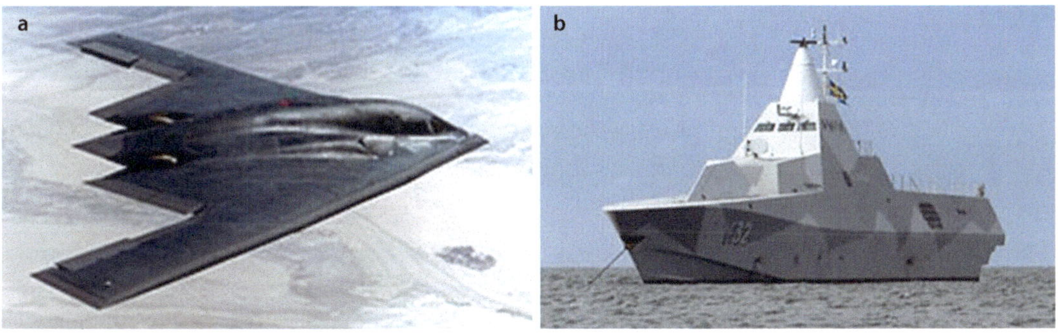

Fig. 2.19 **a** B-2 stealth bomber and **b** Visby class stealth frigate made out of composites [37, 38]

The National Aeronautical Laboratory (NAL, which is an outfit of the Council of Scientific and Industrial Research in India) has developed a CFRP wing using a carbon/epoxy prepreg for 14-seater civil aircraft called Saras. The entire bottom substructure (having dimensions of 5.1 m × 1 m that consist of a skin, 2 spars, 11 ribs, and 8 stringers) has been co-cured in one single operation. The processes used were vacuum bagging and autoclaving.

2.3.3 Stealth Aircrafts and Frigates

Composites play a key role in stealth aircraft and frigates. The B-2 stealth bomber of the US Air Force and the stealth frigate of Sweden (shown in Fig. 2.19) are nearly invisible to radar and are both made out of composites. The B-2 bomber is made out of carbon-reinforced composites, and the matrices used are primarily epoxy and polyimide resins. This composite is able to absorb all radar frequencies, making it completely invisible. In the case of the stealth frigate, it not only absorbs all radar frequencies but also does not leave any magnetic footprint due to the nonmagnetic character of composites. Therefore, it becomes undetectable by enemy radars. Camouflaging paints are applied on its exterior surface to make it visually undetectable too.

2

2.3.4 Automotive and Other Transportation Vehicles

Composites are used in a variety of automotive components, such as exterior body components (hood or door panels), underbody components (radiator supports, bumper reinforcement beams, and door frames), structural body components, chassis components, rear leaf spring, drive shafts, road wheels, engine components, etc. Manufacturers of automotives use composites extensively in order to reduce the weight of the vehicle. Hyundai has developed a CFRP chassis for their car called the Intrado shown in ◘ Fig. 2.20. It has been developed using the VARTM process. The reinforcement used in development of this chassis is a carbon fibre braiding with a closed-cell polyethylene foam as the core material. The matrix used is an epoxy resin. The reduction in weight is 70% compared to a steel chassis. CFRP have also been used in parts of the BMW 7 Series car to reduce its weight.

Low weight is the most desirable property of skybus coaches. Composites are the materials that provide low-weight solutions in load-bearing structures. ◘ Figure 2.21 shows a composite air train.

◘ **Fig. 2.20** CFRP chassis of the Intrado by Hyundai [39]

◘ **Fig. 2.21** Composite air train [40]

2.3.5 Building and Construction

Buildings and their construction are an area that may register the maximum consumption of composites due to the vastness and large volumes of the products (i.e. buildings and other civil structures). The construction industry holds enormous promise in the use of composites in doors and windows, paneling, furniture, nonstructural gratings, long span roof structures, tanks, bridge components, column wraps, cladding, and the repair of concrete, complete bridge systems, and other interiors. A 130-foot-long pedestrian composite bridge was made in Denmark in 1997, and its picture is shown in ◘ Fig. 2.22.

Components made of composite materials find extensive applications in civil structures, door panels, door/window frames, kitchen furniture, etc. with advantages like corrosion resistance, longer life, low maintenance, ease in workability, fire retardancy, etc. Due to the high dielectric property of glass-reinforced composites, they also are being used for the fabrication of high-voltage electrical transmission towers from pultruded composite sections using a 'snap and build' assembly procedure. This process eliminates the use of fasteners and adhesives. The weight of such composite assemblies is less than half that of conventional steel equivalent structures. The composite components can be readily airlifted into remote areas and assembled by small teams. This eliminates the often inaccessibility through roads. In addition, the inherent insulating characteristics of composites permits closer placement of the ceramic holder for the cables, reducing the overall tower size. This leads to general cost reduction.

Partitions, polymer concrete, prefabricated buildings, kiosks, cabins/housing, structural/decorative building elements, bridge elements/sections, furniture, fencing, railings, etc. are more applications of composites in the construction industry.

◘ **Fig. 2.22** A 130-foot-long pedestrian composite bridge [41]

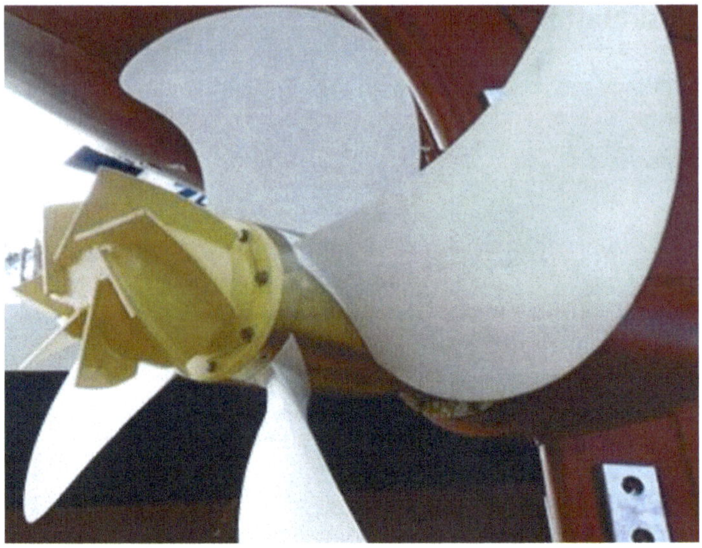

Fig. 2.23 Composite propeller for a merchant vessel [42]

2.3.6 Corrosion Resistance

Chemical plant linings, oil industry components of pipes and ducts, chimneys, grid flooring, staging and walkways, pressure vessels, processing tanks and vessels, fume hoods, scrubbers and cooling tower components, assemblies, enclosures, etc. are often made out of polymeric composites. Resistance to seawater corrosion is an important property of composites. Japan has developed a CFRP propeller for the 499-gigaton merchant vessel shown in Fig. 2.23.

2.3.7 Defence

Airframes, rocket motor liners, satellite components, enclosures and containers, personnel armour, shipping and transit containers, simulators, etc. are made out of composites. Light combat aircraft (LCA) is developed by India and contains more than 40% of the structural components—by weight—made out of composites. The wings, vertical tail, and control surfaces of the LCA are made of CFRP. In an advanced light helicopter (ALH), the nose is made out of Kevlar-reinforced composites and the tail section made out of CFRP. The LCA-Tejas and ALH-Rudra are shown in Fig. 2.24. The rotor hub, cockpit, and main and tail rotor blades are made of carbon fibre-reinforced composites.

The extent of composites used in the structure is about 60% by area and about 29% by weight.

2.3.8 Marine

Composites are abundantly used in the marine industry. The major applications of composites are in hulls of boats and yachts;

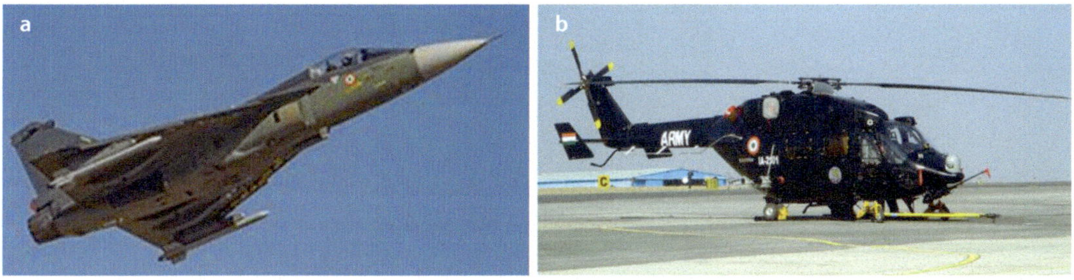

Fig. 2.24 **a** LCA-Tejas and **b** ALH-Rudra [43, 44]

Fig. 2.25 Hull of a high-speed interceptor boat [45]

buoys; boat accessories (such as propeller shafts, masts, bulkheads, rudders, deckhouses, hatch covers, and bow modules); sailboards; mouldings and fittings for ferries and cruise liners; trawlers; etc. Composites exhibit excellent resistance to the marine environment, including salinity, high wind velocity, humidity, and low temperatures.

Complex shapes with monolithic structures are the advantages of hulls. These are the driving factors in the development of composite boats. High-speed boats employ advanced and hybrid composites for both high performance and safety. Aquarius Fibreglass of Goa, India, has manufactured the hull of a high-speed (35 knots) interceptor boat with a length of 50 feet shown in Fig. 2.25. It has a sandwich composite construction.

2.3.9 Electrical and Electronic Appliances

Composites are used for internal and external components and fittings, casings, circuit boards, generation/transmission components, insulators, switch-boxes/cabinets, booms, distribution posts/

Fig. 2.26 Cableway made out of polymeric composites [46]

pylons, transformer elements, ladders, cableways, and many more appliances. Also, the packaging of these items is done quite well using polymeric composites. The eight-seater cableway gondola was built in 2002 of composites in Saint-Martin-de-Belleville, Savoie, France, and is shown in Fig. 2.26.

2.3.10 Sports Equipment

Composites play an important role in the manufacture of sporting goods. The major sporting goods made out of composites are golf shafts, racquets, snowboards, bicycles, baseball bats, hockey sticks, canoes, fishing rods, archery bows, ski poles and skis, surf boards, etc. These all are manufactured by compression moulding, hand layup, prepreg layup, pressure bagging, filament winding, and injection moulding. Tennis racquets are made of composites using spread carbon tows with a reinforced polyurethane matrix. Both tennis racquets and composite bicycles use carbon fibres (Textreme®) as the reinforcements. Some of the sporting goods made of composites are shown in Fig. 2.27.

2.3.11 Healthcare

Composites are used for making fracture braces and orthotic and prosthetic devices for the locomotor disabled. The use of carbon–carbon composites have opened new areas of healthcare applications for the creation of prosthetic tilt disc heart valves, hip joints, and knee joints. There are hoards of support structures made out of composites and being used in medical equipment, such as X-ray machines, dialysis machines, etc. Some examples are given here.

◘ **Fig. 2.27** **a** Carbon composite rafting boat, **b** carbon composite 2-lb bicycle, **c** carbon composite tennis racquet, and **d** golf clubs having carbon composite shafts [47, 48]

- *MRIs*—Composites are made of diamagnetic materials. They don't interfere with a magnetic field; therefore, composites allow the use of large magnets in magnetic resonance imaging (MRI). Composites are used in MRI housing and tables. In addition, composite bars are used in the construction of an MRI room to reinforce the concrete walls and floors.
- *X-Rays*—The X-ray table in an operation theatre is made of composites, preferably carbon fibre-reinforced epoxy resin for taking the X-ray of the repaired part of the body during orthopaedic surgery. Composites are transparent to X-ray. The X-ray machine used in operation theatre is popularly called a 'C arm'. The C-shaped X-ray probe enables technicians to take the X-ray from both the top and the bottom (from under the operation table) as the film is taken or the X-ray result is seen on the monitor.

2.3.12 General Engineering

Composites offer many possibilities in the area of general engineering applications, such as piping systems, air duct systems, storage tanks, pressure vessels, sanitary wares (i.e. sinks and bathtubs), furniture, door panels, door and window frames, etc. The use of composites in these applications is very significant from the point of the life cycle cost of these products. The products made out of composites require quite low maintenance; thus, their life cycle cost is quite low.

Composites and their applications have made a tremendous impact on the world during last three decades or so. The growth of the composite market is above 8% globally. The Asia and Pacific regions are registering a 24% total use of composites. Composites have found wide acceptance due to their ability to adapt to varied environments because they can be made to suit the given environment. Also, being dielectric and noncorrosive in nature, composites can be combined with other materials easily. The world is witnessing new research in the area of composites in terms of materials and processes to make them more cost-effective and versatile.

However, there is a need for an improved awareness, adaptation to technology, versatility, and applications among all stakeholders. The flow of knowledge into industry has not come up to the desired level in many developing and underdeveloped countries. It is high time to bring awareness among industries on technological aspects for the proliferation of technology and subsequent commercialization. The increased awareness and knowledge among industries will significantly boost the use of composites and lead to further innovations in composites.

Fibre-reinforced composites based on a high-performance polymer could provide an enormous scope and opportunity to replace metals. For example, approximately 30% of the components for the super Airbus A380, which will carry 550 passengers, are expected to be made of CFRP.

In next chapter, a detailed look at the micromechanics and macromechanics of polymeric composites will be discussed. This will impart an insight into the mechanism of the engineering behaviour of composites.

To Be Remembered

1. Advantages offered by polymeric composites are in their performance, cost, and production rate.
2. Use of natural fibre reinforcement reduces the weight of composites by 20–30%.
3. Pressure vessels can be made twice as strong in the circumferential direction as in the axial with the use of composites.
4. CFRP leaf springs overcome the need for the maintenance and lubrication of shock absorbers.
5. CFRP leaf springs in a rail wagon has dispensed with the necessity of steel springs.
6. Composites offer the in situ mouldability of metallic inserts.
7. Composites are designed to be the best ablative materials.
8. Composites are good dampening materials and are corrosion resistant.
9. Composites offer radar-transparent properties and are used as nose cones and radomes in aircrafts, missiles, SATCOM antennas, etc.
10. Catastrophic failures occur often in composites.

11. *P*MR-PI composites may provide service temperature in excess of 300 °C.
12. *R*epairing composites is a complex task.
13. *L*ightweight sporting goods (such as racquets, golf clubs, bicycles, boats, etc.) are made of polymeric composites.

Exercises

1. How will you develop a lightweight fishing rod? Explain the reason for your selection of raw materials and processing technologies.
2. What are the reinforcements to be used for underwater applications? Provide their reasons for selection of materials and process too.
3. Why is the manufacture of high-pressure vessels out of composites gaining momentum?
4. In what conditions would a polar wound pressure vessel be made for an automobile? Is there any advantage in using the filament winding process over a tape winding process? Explain.
5. What are the raw materials to be chosen for making the lining of a rocket motor engine?
6. Why are thermoplastic composites used for making fracture braces for patients?
7. Why is a sandwich foamed structure used in boat hulls and car chassis?
8. What are the processes to make a composite cycle frame and an aircraft seat frame?
9. What are the possible advantages in making a cableway cabin out of composites?
10. How do you compare the ability to repair polymeric composites with metals? Which material has better repairability and why?

References

1. Chung, D.D.L., "Structural composite materials tailored for damping," *Journal of Alloys and Compounds*, 355 (2003), p. 216.
2. Maekawa, Z., Hiroyuki, H., and Goto, A., *Nippon Kikai Gakkai Ronbunshu*, 60 (571) (1994), p. 831.
3. Finegan, I.C. and Gibson, R.F., *Compos. Struct.*, 44 (2) (1999), p. 89.
4. Maheri, M.R. and Adams, R.D., *JSME Int. J. Ser. A – Solid Mech. Mater. Eng.*, 42 (3) (1999), p. 307.
5. Tsekmes, I.A., Kochetov, R., Morshuis, P.H.F., and Smit, J.J., "Thermal Conductivity of Polymeric Composites: A Review," *2013 IEEE International Conference on Solid Dielectrics*, Bologna, Italy, June 30–July 4, 2013.
6. Shinde, S.L. and Goela, J.S., *High Thermal Conductivity Materials*, Springer, 2006.
7. Weber, E.H., "Development and modeling of thermally conductive polymer/carbon composites," PhD dissertation, Michigan Technological University, 2001.
8. Theng, B.K., "Interactions of clay minerals with organic polymers. Some practical applications," *Clays and Clay Minerals*, 18 (1970), p. 357.

2

9. Li, A. and Wang, W., "Synthesis and properties of clay-based superabsorbent composite," *Eur. Polym. J.*, 41 (2005), p. 1630.
10. Nickel, L., The sky's limit, *Reinforced Plastics*, 48 (7) July/August 2004, p. 32.
11. Henkle, http://www.henkel-adhesives.com/aerospace/applications-32713.htm?tn=32766.
12. "14 seat civil aircraft, SARAS with the entire wing and empennage developed using composites," *FRP Today*, 13 (7), July 2013, p. 17.
13. "This Old House," http://www.thisoldhouse.com/toh/photos/0,,20337823,00.html.
14. "'New Bus for London' Utilizes Composites to Achieve the Necessary Weight Saving," *FRP Today*, 13 (10), October 2013, p. 18.
15. "Application Trends," *FRP Today News*, 13 (8), August 2013, p. 23.
16. "Vestas produced most powerful blade in wind industry," *FRP Today*, 13 (10), October 2013, p. 18.
17. 'Divinycell makes EzyTopper super light and super fast', FRP Today, 15 (1), January 2015, p. 36.
18. "The High Curvature Armour System," *FRP Today*, 13 (7), July 2013, p. 25.
19. Krishnan, M., "Biocomposites in Automotive Applications," *FRP Today News*, 13 (8), August 2013, p. 21.
20. "efWING- The new railway vehicle truck having CFRP frame wins award," *FRP Today*, 14 (5), May 2014, p. 24.
21. "World's First Rolling Stock Bogie with a CFRP Frame," *FRP Today*, 13 (8), August 2013, p. 18.
22. "Composite leaf spring for new Volvo SUV XC 90," *FRP Today*, 15 (1), January 2015, p. 31.
23. Matweb Data Sheet, 2016.
24. "Future Fibres Completes Work on 76 Meter Mast for Perini Navi's Latest Superyacht," *FRP Today*, 13 (10) (2013), p.19.
25. Ran, Z., Yan, Y., Li, J., Qi, Z. and Yang, L., "Determination of thermal expansion coefficients for unidirectional fiber-reinforced composites," *Chinese Journal of Aeronautics*, Volume 27, Issue 5 (2014), p. 1180.
26. Bhandari, Bhaumik A. and Patel, Bhavesh C., "Parametric Analysis Of Composite Leaf Spring," *IJSTE–International Journal of Science Technology & Engineering*, Vol.1, Issue 1 (2014) ISSN(online), p. 2349.
27. Sangale, S.S., Kale, K.B., and Dighe, Y.S., "Design Analysis of Carbon/Epoxy Composite LeafSpring," *International Journal of Research in Advent Technology* (E-ISSN: 2321–9637) *Special Issue1st International Conference on Advent Trends in Engineering, Science and Technology*- ICATEST 2015, 08 March 2015, p. 172.
28. *The Drone race: DRDO Rustom II, Monitoring the 21st Century Asian Arms Race*, https://21stcenturyasianarmsrace.files.wordpress.com/2015/07/indian-rustom-2-male-uav1.jpg, 2 July 2015.
29. "Boeing 777X to Deliver Unprecedented Efficiency and Economics," *FRP Today*, 13 (12), December 2013, p. 27.
30. "Automobile natural Gas tanks Take the Pressure with Thermoplastic Composites," *FRP Today*, 14 (11), November 2014, p. 34.
31. *American Airlines Flight 587*, https://en.wikipedia.org/wiki/American_Airlines_Flight_587.
32. *Composites repairs*, https://en.wikipedia.org/wiki/Composite_repairs.
33. Valavan, M.T., *Adverse impact of composite material, America's Cup Boat 'Young America*, https://www.slideshare.net/manothirumal/composites-47072751.
34. *Greenhouse*, https://en.wikipedia.org/wiki/Greenhouse.
35. *Boeing 787 Dreamliner*, https://en.wikipedia.org/wiki/Boeing_787_Dreamliner.
36. "'Robo-Raven'- a major Breakthrough for Micro Air Vehicle," *FRP Today*, 13 (7), July 2013, p. 15.
37. *Northrop Grumman B-2 Spirit*, Wikipedia.
38. *Visby class corvette*, Wikipedia.
39. "Hyundai Intrado Car with CFRP Chassis Technology Uses Scott Bader Advanced Materials," *FRP Today*, 15 (5), May 2015, p. 28.

40. "Brazil installs 100% composite air train in preparation for world cup," *FRP Today News*, 14 (1), January 2014, p. 22.
41. *Future bridges of reinforced plastic*, http://www.netpublikationer.dk/um/9563/html/printerversion_entire_publication.htm.
42. "World's First Installation of a CFRP Propeller on a Merchant Vessel," *FRP Today*, 14 (9), September 2014, p. 30.
43. HAL- Tejas, Wikipedia.
44. HAL- Rudra, Wikipedia.
45. "CCG and Acquarius Fibreglas Develop the First fully Infused Boat Hull in India," *FRP Today*, 15 (4), April 2015, p. 34.
46. *Gondolas*, Poma, Wikipedia.
47. "Berria Bike Launches Frames Reinforced by Textreme," *FRP Today*, 14 (4), April 2014, p. 31.
48. "Carbon fibre in Prince Tennis racquet," *FRP Today*, 15 (5), May 2015, p. 27.

Micromechanics and Macromechanics of Polymeric Composites

The original version of this chapter was revised with the late correction received from the author. The correction to this chapter can be found at https://doi.org/10.1007/978-81-322-3933-8_10

A. K. Ghosh, M. Dwivedi, *Processability of Polymeric Composites*,
https://doi.org/10.1007/978-81-322-3933-8_3

3

Mechanics of composites is an important study in the understanding of the mechanical responses and their mechanisms due to anisotropy. Micromechanics and macromechanics of composites deal with a material at its constituent and bulk levels, respectively. The understanding of the properties of the constituents—their interaction and bonding mechanisms—plays a vital role in understanding micromechanics. On the other hand, macromechanics primarily deals with engineering aspects of composite material and its responses to applied loads. This chapter explains in detail the concepts of micromechanics and macromechanics.

3.1 Micromechanics of Polymeric Composites

Micromechanics is the study and analysis of composite materials at the level of the individual constituents. It also can be said that micromechanics is the study of the behaviour of fibres and filaments, matrices, interfaces, and interphases in a composite material under load. It discusses the phenomenon of the interaction of fibres with the matrix, the role of the interface, the role of the interphase, and the interaction between fibres through the matrix under the influence of external forces.

3.1.1 Rule of Mixtures

Some of the properties of a composite material can be expressed as the sum of the product of the volume fraction of the constituents and their properties. The *volume fraction* of a constituent in a composite material is defined as the ratio of the volume of the constituent to the total volume of the composite material. The resulting expression is commonly referred to as the *rule of mixtures* [1]. In order to achieve optimum properties in advanced composite materials as per the rule of mixtures, the volume fractions of reinforcement and matrix should be 0.5 for each. For example, the tensile strength of the fibre-reinforced composite may be calculated as

$$\sigma_C = \sigma_f V_f + \sigma_m V_m \tag{3.1}$$

where

- σ_c = Tensile strength of composite
- σ_f = Tensile strength of fibre
- σ_m = Tensile strength of matrix
- V_f = Volume fraction of fibre
- V_m = Volume fraction of matrix

Similarly, other properties of the composites may be evaluated theoretically using the rule of mixture. The longitudinal modulus is given as

$$E_c = E_f V_f + E_m V_m \tag{3.2}$$

or

$$E_c = E_f V_f + E_m (1 - V_f)$$

where
- E_c = Young's modulus of composite
- E_f = Young's modulus of fibre
- E_m = Young's modulus of matrix

The transverse modulus, E_T, may be given as

$$\frac{1}{E_T} = \frac{V_f}{E_f} + \frac{V_m}{E_m} \quad (3.3)$$

The shear modulus of composite may be given as

$$G_c = \frac{G_f G_m}{G_f V_m + G_m V_f} \quad (3.4)$$

where
- G_c = Shear modulus of composite
- G_f = Shear modulus of fibre
- G_m = Shear modulus of matrix

However, the theoretically calculated values are higher than that of the experimental values. The reasons to this discrepancy are many. The most important reasons are:
(a) Fibres of nonuniform strength
(b) Discontinuity in the fibres
(c) Interfacial conditions
(d) Disorientation of fibres
(e) Residual stresses
(f) Presence of voids

Interfacial conditions may deteriorate due to either mishandling or improper storage of the fibres. Nonuniformity of strength and discontinuity of fibres occur due to mishandling of the fibres. The role of the interface and the interphase becomes very important in deciding the overall properties of composites. The molecular structure and surface treatments on both reinforcement and matrix define the compatibility of these interfaces, thus the properties of interphase. The disorientation of fibres, residual stress, and the presence of voids occur during processing of composites. The volume fraction of voids may be calculated as

Point to Ponder...
It is virtually impossible to make composites free from voids or defects. In metals also, defects in the form of slip or screw dislocations are present, but they make machinability of metals possible.

$$V_V = \frac{\rho_{ct} - \rho_{ce}}{\rho_{ct}} \quad (3.5)$$

where
- Vv = Volume fraction of voids
- ρ_{ct} = Theoretical density of composite
- ρ_{ce} = Experimentally determined density of composite

Void fraction influences the properties of composites. Higher void fraction leads to lower mechanical properties of composites. The void fraction is linked with processability of composites. The parameters of reinforcement and matrix coupled with processing technology determine the void fraction. For example, filament wound composites having a low viscosity resin with no residue during curing will result in a low void fraction and compact composite [2, 3].

Example 3.1 In a carbon fibre-reinforced epoxy composite, the densities of carbon fibres and matrix are 1.8 g/cm^3 and 1.2 g/cm^3, respectively. It has volume fractions of reinforcement and matrix at 0.5 each. The density of this composite measured is 1.4 g/cm^3. Calculate the void contents in this composite.

Answer

$$\rho_{ce} = 1.4 \text{ g/cm}^3$$

$$\rho_f = 1.8 \text{ g/cm}^3$$

$$\rho_m = 1.2 \text{ g/cm}^3$$

$$V_f = 0.5$$

$$V_m = 0.5$$

$$\rho_{ct} = V_f \rho_f + V_m \rho_m \quad \rho_{ct} = 1.5 \text{ g/cm}^3$$

$$V_v = \frac{\rho_{ct} - \rho_{ce}}{\rho_{ct}} \quad V_v = 0.06$$

For good composites, the strain at a break for the matrix must be higher than the strain at a break for fibre,

$$\varepsilon_m > \varepsilon_f \tag{3.6}$$

where

- ε_m = Elongation at break for matrix
- ε_f = Elongation at break for fibre

$\varepsilon_m > \varepsilon_f$ ensures that the load is borne by the fibres, and the full strength of the fibres is utilized. At very low fibre–volume fractions, the fibres act as cylindrical voids, and as a result, the strength of the composite material is actually lower than that of the unreinforced matrix. In this connection, a parameter called the *critical fibre-volume fraction*, V_{fcrit}, can be introduced (Fig. 3.1). Note that the volume fraction of fibre must be more than the critical fibre volume fraction if fibres have to contribute to the composite strength. Another term, *minimum fibre volume fraction*, V_{fmin}, is defined as the fibre volume fraction below which the addition of fibre reduces the composite strength.

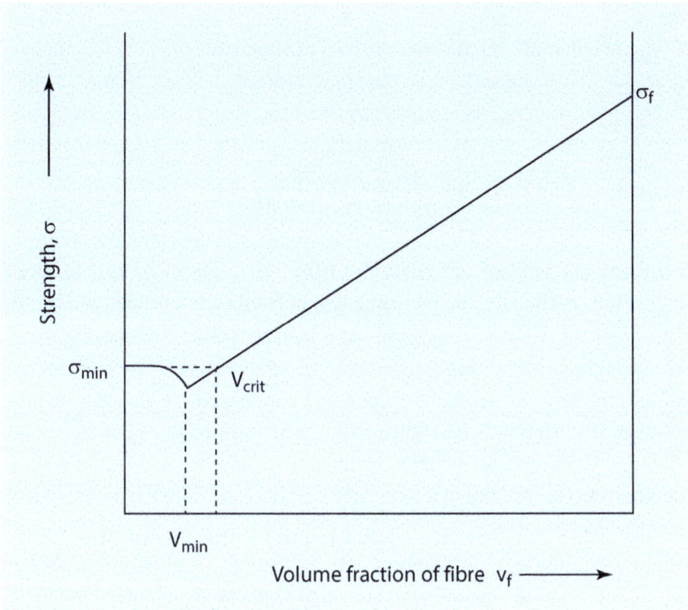

Fig. 3.1 Critical volume fraction of fibre and corresponding strength

The value of V_{crit} is given as

$$V_{crit} = \frac{\sigma_{mu} - \sigma_m^*}{\sigma_{fu} - \sigma_m^*} \tag{3.7}$$

where
- σ_{mu} = Ultimate tensile strength of the matrix
- σ_m* = Tensile strength of the matrix at fracture
- σ_{fu} = Ultimate tensile strength of fibre

The ultimate tensile strength of a composite with V_{crit} may be given as

$$\sigma_{cu} = \sigma_{mu} = \sigma_{fu} V_{crit} + \sigma_m^* \left(1 - V_{crit}\right) \tag{3.8}$$

At V_{min}, the composite will not fail at a stress predicted at Eq. (3.7). At V_{min}, the volume fraction of fibres is ineffective in restraining the elongation of the matrix; therefore, the fibres are strained to fracture. So,

$$V_{min} = \frac{\sigma_{mu} - \sigma_m^*}{\sigma_{fu} + \sigma_{mu} - \sigma_m^*} \tag{3.9}$$

The ultimate tensile strength of composite may be given as

$$\sigma_{cu} = \sigma_{mu} \left(1 - V_{min}\right) \tag{3.10}$$

$$= \sigma_{fu} V_{min} + \sigma_m^* \left(1 - V_{min}\right) \tag{3.11}$$

3

where
- σ_{cu} = Ultimate tensile strength of composite
- σ_{mu} = Ultimate tensile strength of matrix
- σ_{fu} = Ultimate tensile strength of fibre
- $\sigma*_{m}$ = Tensile strength of matrix at fracture
- V_{crit} = Critical volume fraction of fibre
- V_{min} = Minimum volume fraction of fibre

If the matrix is stronger than the fibre, V_{min} and V_{crit} will be very large. Hence, the resultant composite is strong. If a strain at a break of the matrix is equal to a strain at a break of the fibre (i.e. $\varepsilon_m = \varepsilon_f$), the strength of the composite would be reduced, and the full potential of the fibre strength would not be realized. If the strain at a break of the matrix is less than the strain at a break of the fibre (i.e. $\varepsilon_m < \varepsilon_f$), the strength of the composite would be reduced further, and the matrix will fail at a very low stress [1, 4]. The load would not be borne by the fibres, and the full potential of fibre strength would not be realized. The composite will manifest the properties of the matrix only. This will lead to the initiation of cracks and areas of stress concentration. The different modes of failure in composites are discussed in the following sections.

Example 3.2 The strengths of a Kevlar fibre and matrix are 2350 MPa and 100 MPa, respectively. The experimental tensile strength of this composite is 400 MPa. It has volume fractions for the reinforcement and matrix 0.4 and 0.6, respectively. (a) Calculate the critical volume fraction, minimum volume fraction, and void contents of this composite. (b) Estimate the possible nomenclature of the matrix.

Answer

(a)

$\sigma*_{c}$ = 600 MPa

σ_{fu} = 2350 MPa

σ_{mu} = 100 MPa

$V_f = 0.4$

$V_m = 0.6$

$$\sigma_c^* = V_f\sigma_f^* + V_m\sigma_m^*$$

$$\sigma_m^* = \frac{\sigma_c^* - V_f\sigma_f}{V_m};\ \sigma_m^* = 90\ \text{MPa}$$

$$V_{crit} = \frac{\sigma_{mu} - \sigma_m^*}{\sigma_{fu} - \sigma_m^*};\ V_{crit} = 0.004$$

$$V_{min} = \frac{\sigma_{mu} - \sigma_m^*}{\sigma_{fu} + \sigma_{mu} - \sigma_m^*};\ V_{min} = 0.004$$

(b) This matrix is expected to be a thermoplastic matrix such as PEI or PEEK.

3.1.2 Mechanism of Load Transfer

Polymeric composites have maximum properties in the direction of reinforcement. Therefore, unidirectional composites have maximum mechanical properties in the longitudinal direction along the fibres. Optimum volume fraction of fibres leads to the best performance of composites. The optimum volume fraction for the fibres and matrix of composites are 0.5 each. In composites, a load is transferred from fibre to fibre through the matrix by shear force, as shown in ◘ Fig. 3.2. The load is transferred through the fibre ends and small fibre lengths near the fibre ends in short fibre composites.

When the length of the fibre is greater than the length over which the transfer of stress takes place, the end effect is neglected, and the stress can be assumed to be over the entire length of continuous fibre.

In the case of the deformation of a composite on the application of a load, four stages are considered:

(i) Both fibre and matrix deform elastically.
(ii) Fibres deform elastically, but the matrix deforms plastically.

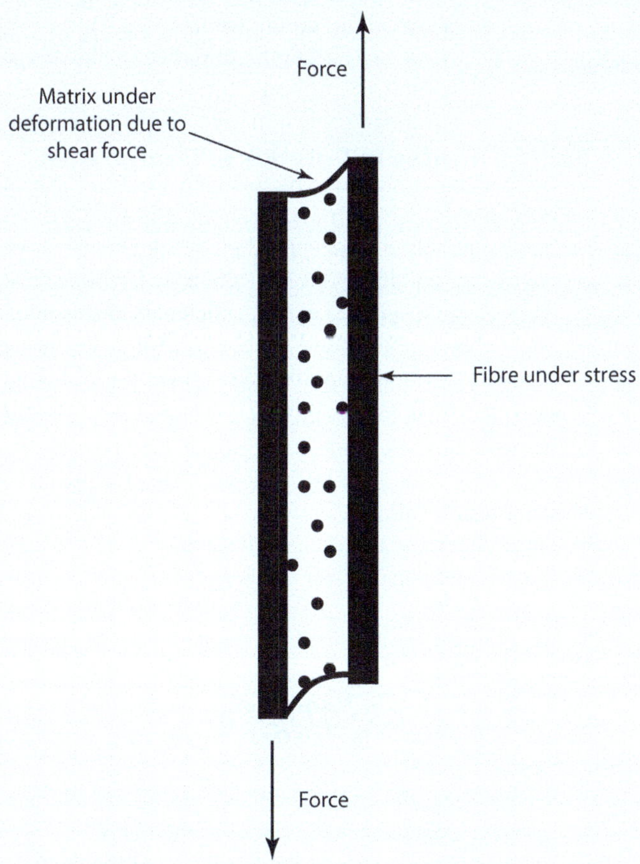

◘ **Fig. 3.2** Fibre-to-fibre load transfer through the matrix

(iii) Both fibres and the matrix deform plastically.
(iv) The fracture of fibres is followed by the fracture of a composite.

The loss of properties in composites is due to the following factors in general.
(i) Stress corrosion causing a loss of strength of fibres
(ii) Chemical degradation of the matrix
(iii) Loss of interfacial strength between the fibres and the matrix
(iv) Degradation of the matrix due to overaging, UV effects, service temperature, and other radiations (such as X-ray)

In general, brittle failure is observed in polymeric composites due to a very low percentage of elongation at the break, as mentioned in ▶ Table 1.6 of ▶ Chap. 1. The strengthening effect is primarily achieved by the fibres in the composite, and it is only achieved when the ultimate tensile strength of the composite exceeds that of the matrix alone. In UD composites, strength in the transverse direction is lower than the strength of the matrix due to stress and strain caused by the fibres in the matrix. The stress and strain concentration causes failure of composites at a much lower strain than the strain at which an unstrained matrix would fail. However, the modulus in the transverse direction is more than that of the matrix if the volume fraction of fibres is very high and the fibres are of a high modulus.

3.1.3 Failure Modes in Polymeric Composites

The failure in composites is a distinct phenomenon. It initiates from microcrazing of the matrix and leads to surface and core cracks in the matrix. Furthermore, it causes debonding between the fibre and the matrix, leading to breaking of the fibres [5, 6]. Finally, delamination of the composite takes place. Composites with low V_f (<0.40), moderate V_f (between 0.40 and 0.65), and high V_f (>0.65) typically show a brittle failure with initial cracking leading to debonding and brittle failure with fibre pull-out, respectively. The possible three modes of failure under longitudinal compressive load are given here:
(i) Transverse tensile failure
(ii) Fibres microbuckling in extensional mode
(iii) Fibres microbuckling in shear mode

Similarly, the possible three modes of failure under longitudinal tensile load are as follows:
(i) Tensile forces stretch the matrix more than the fibres, causing the material to shear at the interface between the matrix and the fibres.
(ii) Tensile forces near the end of the fibres exceed the tolerances of the matrix, separating the fibres from the matrix
(iii) Tensile forces can also exceed the tolerances of the fibres causing the fibres themselves to fracture, leading to material failure

The energy required per unit area of composite for the fracture of the fibres in tension mode is given as

$$\text{Energy} = \frac{V_f \sigma_{fu}^2 l}{6E_f} \tag{3.12}$$

where

- σ_{fu} = Ultimate tensile strength of fibre
- l = Fibre length
- E_f = Young's modulus of fibre

The composite may fail due to either debonding of the fibre with the matrix and/or by fibre pull-out. *Fibre pull-out* occurs in brittle or short fibre polymeric composites, and the fibre pull-out energy per unit area is given as

$$\text{Fiber pull} - \text{out energy} = \frac{V_f \sigma_{fu} l_c}{12} \tag{3.13}$$

where

- σ_{fu} = Ultimate tensile strength of fibre
- l_c = Critical length

Similarly, the energy required for the fracture in the matrix per unit area of composite is given as

$$\text{Energy} = \frac{(1-V_f)^2 \sigma_{mu} d U_m}{V_f 4\tau} \tag{3.14}$$

where

- σ_{mu} = Ultimate tensile strength of the matrix
- d = Fibre diameter
- U_m = Work done to deform the matrix to rupture per unit volume (U_m is negligible in brittle matrix composites and, therefore, may be ignored.)
- τ = Interfacial shear stress

The coefficient of thermal expansion of UD composites is lower in the longitudinal direction than that of the transverse direction. The coefficient of thermal expansion of fibres is less than that of the matrix. Therefore, fibres in the longitudinal direction put the restrain in the longitudinal coefficient of thermal expansion. The transverse coefficient of expansion of UD composites is large at low V_f and can be larger than that of unreinforced matrix because the matrix that is restrained to expand in longitudinal direction due to fibres is forced to expand in transverse direction [5].

Example 3.3 Find the energy required for fibre fracture and pulling out the fibre for a glass fibre-reinforced composite.The tensile strength of fibres, Young's modulus of fibres, length of fibre, diameter of glass fibre, and yield stress of the matrix in shear are 3500 MPa, 70 GPa, 100 mm, 15 μm, and 24 MPa, respectively. The composite has volume fractions of reinforcement and a matrix at 0.5 each. Calculate the critical aspect ratio of fibre in this composite.

3

Answer

$\sigma_{fu} = 3500$ MPa

$E_f = 70$G Pa

$l = 100$ mm

$d = 15$ μm

$\tau_y = 24$ MPa

$V_f = 0.5$

$V_m = 0.5$

$$\frac{l_c}{d} = \frac{\sigma_{fu}}{2\tau_y} \; l_c = d\frac{\sigma_{fu}}{2\tau_y} \; l_c = 2.19 \text{ mm}$$

$$\text{Energy} = \frac{V_f \sigma_{ful}^2}{6E_f}$$
$$= 145.8 \times 10^3 \text{ J}$$

$$\text{Fiber pull-out energy} = \frac{V_f \sigma_{fu} l_c}{12}$$
$$= 319.38 \times 10^2 \text{ J}$$

3.1.4 Short Fibre Composites

In short fibre composites, the end effect of the fibres are not neglected as is the case in long fibre composites. In short fibre-reinforced composites, the ends of the fibres and region near the ends bear the maximum stress. The maximum stress occurs at the mid-fibre length. The load transfer length of the fibre for the applied stress is a critical length for the fibre (l_c). The l_c is calculated as

$$\frac{l_c}{d} = \frac{\sigma_{fu}}{2\tau_y} \tag{3.15}$$

where
- l_c = Critical length for fibre
- σ_{fu} = Ultimate tensile strength of fibre
- d = Diameter of fibre
- τ_y = Yield stress of matrix in shear

The yield stress of the matrix in shear (τ_y) is equal to interface shear stress along the fibre length. It can be given as

$$\sigma_{fu} = \frac{2\tau_y Z}{r} \tag{3.16}$$

where
- Z = Mid-fibre length (i.e. $l/2$ where l = fibre length of short fibre)
- r = Fibre's radius

If the length of the fibre is smaller than l_c, the stress borne by the fibre would be less than its actual stress bearing capacity. If length is equal to l_c, the entire length of the fibre would bear the stress, and the maximum stress would be at the centre of the fibre. If the length of fibre is more than l_c, the composite will behave as a continuous fibre-reinforced composite. When the force is applied on a short fibre composite, the shear force is at a maximum at the ends of the fibres and zero at the centre of the fibre [1, 7, 8]. However, the tensile force is at a maximum at the centre of fibre and zero at the ends of fibre.The strength of short fibre-reinforced composites would follow the rule of mixtures as described in Eq. (3.1) if the fibre length $l >> l_c$. In the case where $l < l_c$ and $l = l_c$, the *ultimate tensile strength* of a composite is given as

$$\sigma_{cu} = \frac{\tau_y l}{d} V_f + \sigma_{mu} V_m \quad \text{for}\, l > l_c \tag{3.17}$$

$$\sigma_{cu} = \sigma_{fu}\left(1 - \frac{l_c}{2l}\right) V_f + (\sigma_m)\varepsilon_f^* V_m \quad \text{for}\, l < l_c \tag{3.18}$$

where
- σ_{cu} = Ultimate tensile strength of composite
- σ_{mu} = Ultimate tensile strength of matrix
- σ_{fu} = Ultimate tensile strength of fibre
- σ_m = Tensile strength of fibre
- l_c = Critical length for fibre
- d = Diameter of fibre
- τ_y = Yield stress of matrix in shear
- V_f = Volume fraction of fibre
- ε^*_f = Fibre fracture strain

The modulus of the radome fibre composite may be calculated as

$$E_{Random} = \left(\frac{3}{8}\right) E_L + \left(\frac{5}{8}\right) E_T \tag{3.19}$$

where
- E_L = Longitudinal modulus
- E_T = Transverse modulus

The Halpin-Tsai equation for longitudinal and transverse moduli of short fibre-reinforced composite is given as

$$\frac{E_L}{E_m} = \frac{1 + \frac{2l}{d}\eta_L V_f}{1 - \eta_L V_f} \tag{3.20}$$

3

$$\frac{E_L}{E_m} = \frac{1+2\eta_T V_f}{1-\eta_T V_f} \tag{3.21}$$

where

- E_L = Longitudinal modulus of short fibre-reinforced composite
- E_m = Modulus of matrix
- E_f = Modulus of fibre
- l = Length of fibre
- d = Diameter of fibre
- η_L = Longitudinal coefficient of fibre
- η_T = Transverse coefficient of fibre
- V_f = Volume fraction of fibre

The coefficients of fibre, η_L and η_T, are given as

$$\eta_L = \frac{(E_f / E_m)-1}{(E_f / E_m)+2(l/d)} \tag{3.22}$$

$$\eta_T = \frac{(E_f / E_m)-1}{(E_f / E_m)+2} \tag{3.23}$$

Nevertheless, it is important to appreciate under what service conditions advanced composites and structures operate. These conditions demand the design of composites must suit to a loading condition. The principles and equations mentioned herein only pave the way for such designs and provide a broader understanding of the response of advanced composites.

3.2 Macromechanics of Polymeric Composites

Macromechanics is the study of stresses and strains at the macro level in a material. In composites, macromechanics is the study of the gross behaviour of a lamina or a laminate when forces are applied on it. A single lamina is generally orthotropic, which has three mutually perpendicular planes of symmetry and is often transversely isotropic.

3.2.1 Analysis of Anisotropic Composite Laminae

In order to understand the macromechanics of anisotropic composite laminae, it is essential to understand three-dimensional (3-D) distribution of stresses on a unit element like a cube. In a cube, the edges are parallel to the coordinate axes *x*, *y*, and *z*. The forces are acting on three perpendicular faces of the cube, and the forces are resolved into nine stress components, as shown in ◘ Fig. 3.3. The stress at point in a body is defined by nine components of the stress tensor σ_{ij}. A *tensor* is a mathematical or physical expression that transforms according to a specific law of transformation with a change in the coordinate system. A *stress*

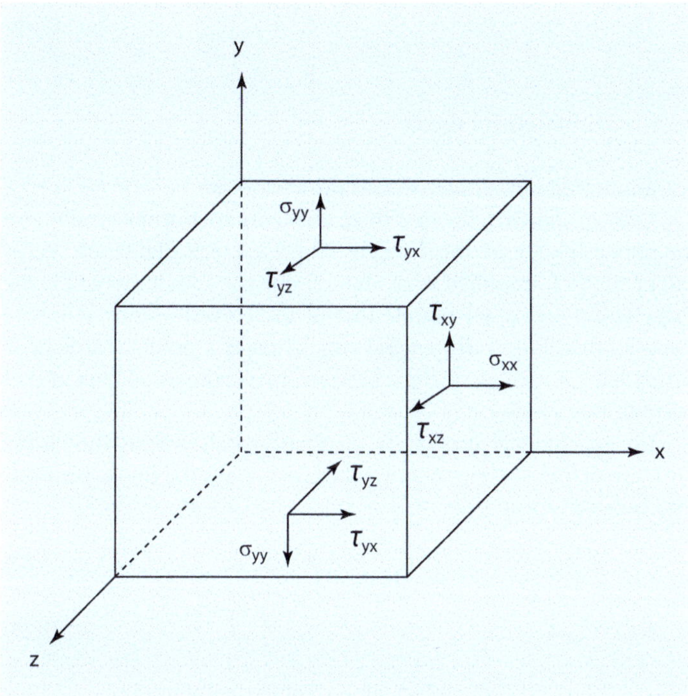

Fig. 3.3 Component of tensile stresses and torque acting on a unit cube

tensor is a convenient way to express the state of stress at a point. It is a second-order Cartesian tensor in which the three diagonal elements represent normal stresses in three mutually orthogonal directions and the remaining six elements represent the shear stresses [9, 10]. As per standard notation, a stress σ_{ij} acts in j directionon a plane normal to the i direction. If $i = j$, σ_{ij} is a normal stress, and if $i \neq j$, σ_{ij} is a shear stress. The stress, for example, σ_{xx} refers to the tensile stress in plane normal to x direction. In the absence of any internal torque of the cubic body, the total torque acting on a cubic body must be zero; therefore, $\tau_{xy} = \tau_{yx}$, $\tau_{xz} = \tau_{zx}$, and $\tau_{yz} = \tau_{zy}$. This implies that there are only six independent stresses that consist of three normal stresses and three shear stresses. The stress tensor may be denoted as σ_{ij}, and its six components may be written as

$$\sigma_{ij} = \begin{pmatrix} \sigma_{xx} & \sigma_{xy} & \sigma_{xz} \\ \sigma_{yx} & \sigma_{yy} & \sigma_{yz} \\ \sigma_{zx} & \sigma_{zy} & \sigma_{zz} \end{pmatrix} \tag{3.24}$$

The corresponding strain tensor may be written as

$$\varepsilon_{ij} = \begin{pmatrix} \varepsilon_{xx} & \varepsilon_{xy} & \varepsilon_{xz} \\ \varepsilon_{yx} & \varepsilon_{yy} & \varepsilon_{yz} \\ \varepsilon_{zx} & \varepsilon_{zy} & \varepsilon_{zz} \end{pmatrix} \tag{3.25}$$

However, it is to be noted that for $i \neq j$, the engineering shear strain γ_{ij} differs from tensorial strain ε_{ij}, and their relationship is given as

3

$$Y_{ij} = 2\varepsilon_i = e_{ij} + e_{ji} \tag{3.26}$$

where

- e_{ij} = Displacement vector

(In an alternate notation, the strain tensor can be expressed as a 9 × 1 vector in which the first three elements are normal strains and the remaining six are shear strains. The 9 × 1 tensorial strain vector can be further simplified as an engineering 6 × 1 vector in which the shear strain terms are double that of the corresponding tensorial shear strains. Physically, *engineering strain* at a point in a plane is defined as the change in angle between two orthogonal lines at that point in that plane.)

The mechanical properties of elastic solids within the elastic limit are defined by *Hook's law*, which expresses the linear relationship between σ and ε as

$$\varepsilon_{ij} = S_{ijkl}\sigma_{kl} \tag{3.27}$$

$$\sigma_{ij} = C_{ijkl}\varepsilon_{kl} \tag{3.28}$$

where S_{ijkl} and C_{ijkl} are compliance and stiffness tensors, respectively, and

$$\varepsilon_{ij} = \varepsilon_{xx}, \varepsilon_{yy}, \varepsilon_{zz}, \ldots, \text{etc.} \tag{3.29}$$

$$\sigma_{kl} = \sigma_{xx}, \sigma_{yy}, \sigma_{zz}, \ldots, \text{etc.} \tag{3.30}$$

The *i*, *j*, *k*, and *l* of *S* and *C* are assigned values 1, 2, and 3, and this becomes synonymous to *x*, *y*, and *z*, respectively. In general, 1, 2, and 3 and *x*, *y*, and *z* represent local and global coordinates, respectively. It is more convenient to work with a compliance constant rather than with a stiffness constant due to the ease in experimental procedures. Therefore, strain tensor ε_{xx} and other strain components, such as ε_{yy}, γ_{xy}, etc., may be expressed as follows:

$$\begin{aligned}\varepsilon_{xx} = {} & S_{1111}\sigma_{xx} + S_{1122}\sigma_{yy} + S_{1133}\sigma_{zz} + S_{1113}\sigma_{XZ} \\ & + S_{1123}\sigma_{YZ} + S_{1112}\sigma_{XY}\end{aligned} \tag{3.31}$$

$$\begin{Bmatrix} \varepsilon_{xx} \\ \varepsilon_{yy} \\ \varepsilon_{zz} \\ \gamma_{yz} \\ \gamma_{zx} \\ \gamma_{xy} \end{Bmatrix} = \begin{bmatrix} S_{1111} & S_{1122} & S_{1133} & S_{1123} & S_{1131} & S_{1112} \\ S_{2211} & S_{2222} & S_{2233} & S_{2223} & S_{2231} & S_{2212} \\ S_{3311} & S_{3322} & S_{3333} & S_{3323} & S_{3331} & S_{3312} \\ S_{2311} & S_{2322} & S_{2333} & S_{2323} & S_{2331} & S_{2312} \\ S_{3111} & S_{3122} & S_{3133} & S_{3123} & S_{3131} & S_{3112} \\ S_{1211} & S_{1222} & S_{1233} & S_{1223} & S_{1231} & S_{1212} \end{bmatrix} \begin{Bmatrix} \sigma_{xx} \\ \sigma_{yy} \\ \sigma_{zz} \\ \tau_{yz} \\ \tau_{zx} \\ \tau_{xy} \end{Bmatrix} \tag{3.32}$$

This is written in abbreviated form as

$$\varepsilon_p = \sum S_{pq}\sigma_q \tag{3.33}$$

$$\sigma_p = \sum C_{pq}\varepsilon_q \tag{3.34}$$

where ε_p and σ_p represent six independent components of strain and stress, respectively. p and q run from 1 to 6. This can be shown by converting ij to p and kl to q according to the following scheme.

Tensor notation 11 22 33 23 32 13 31 12 21

Matrix notation 1 2 3 4 5 6

Therefore, the first line of Eq. 3.28 in the contracted notation may be written as

$$\sigma_{xx} = C_{11}\varepsilon_{xx} + C_{12}\varepsilon_{yy} + C_{13}\varepsilon_{zz} + C_{14}\varepsilon_{XZ} + C_{15}\varepsilon_{YZ} + C_{16}\varepsilon_{XY} \tag{3.35}$$

In line with Eq. 3.35, expressions for other stress components can be derived at in terms of the stiffness matrix terms and strains. Thus,

$$\begin{pmatrix} \sigma_1 \\ \sigma_2 \\ \sigma_3 \\ \tau_1 \\ \tau_2 \\ \tau_3 \end{pmatrix} = \begin{pmatrix} C_{11} & C_{12} & C_{13} & C_{14} & C_{15} & C_{16} \\ C_{12} & C_{22} & C_{23} & C_{24} & C_{25} & C_{26} \\ C_{13} & C_{23} & C_{33} & C_{34} & C_{35} & C_{36} \\ C_{14} & C_{24} & C_{34} & C_{44} & C_{45} & C_{46} \\ C_{15} & C_{25} & C_{35} & C_{45} & C_{55} & C_{56} \\ C_{16} & C_{26} & C_{36} & C_{46} & C_{56} & C_{66} \end{pmatrix} \begin{pmatrix} \varepsilon_1 \\ \varepsilon_2 \\ \varepsilon_3 \\ Y_1 \\ Y_2 \\ Y_3 \end{pmatrix} \tag{3.36}$$

3.2.2 Orthotropy in Composites

A composite lamina is orthotropic in nature (i.e. it has symmetrical structure in three principal planes). Its elastic properties are defined by nine independent elastic constants. The thickness of lamina is much smaller than other two dimensions (i.e. length and width). For simplicity, the stresses normal to the plane of the lamina are ignored. This is commonly referred to as the plane stress condition (i.e. $\sigma_3 = \sigma_{13} = \sigma_{23} = 0$). However, strains are present when normal to the lamina. The stress-strain relation can be written as

$$\begin{pmatrix} \sigma_1 \\ \sigma_2 \\ \tau_{12} \end{pmatrix} = \begin{pmatrix} C_{11} & C_{12} & 0 \\ C_{12} & C_{22} & 0 \\ 0 & 0 & C_{66} \end{pmatrix} \begin{pmatrix} \varepsilon_1 \\ \varepsilon_2 \\ \tau_{12} \end{pmatrix} \tag{3.37}$$

where four independent elastic constants (modulus and Poisson's ratio) of lamina are E_1, E_2, ν_{12}, and ν_{21} are given as

$$C_{11} = \frac{E_1}{1 - \nu_{12}\nu_{21}} \tag{3.38}$$

$$C_{12} = \frac{\nu_{12}E_2}{1-\nu_{12}\nu_{21}} = \frac{\nu_{21} / E_2}{1-\nu_{12}\nu_{21}} \tag{3.39}$$

$$C_{22} = \frac{E_2}{1-\nu_{12}\nu_{21}} \tag{3.40}$$

$$C_{66} = G_{12} \tag{3.41}$$

Similar consideration may lead Eq. 3.33 to be

$$\begin{pmatrix} \varepsilon_1 \\ \varepsilon_2 \\ Y_{12} \end{pmatrix} = \begin{pmatrix} S_{11} & S_{12} & 0 \\ S_{12} & S_{22} & 0 \\ 0 & 0 & S_{66} \end{pmatrix} \begin{pmatrix} \sigma_1 \\ \sigma_2 \\ \tau_{12} \end{pmatrix} \tag{3.42}$$

Here four independent elastic constants (modulus and Poisson's ratio) of lamina are E_1, E_2, ν_{12}, and ν_{21} are given as

$$S_{11} = 1 / E_1 \tag{3.43}$$

$$S_{22} = 1 / E_2 \tag{3.44}$$

$$S_{66} = 1 / G_{12} \tag{3.45}$$

$$S_{12} = -\nu_{12} / E_1 = -\nu_{21} / E_2 \tag{3.46}$$

Strengths and the modulus of elasticity are related to the processability of polymeric composites as

$$\sigma_{11} \text{ or } E_1 = MG_{12} \tag{3.47}$$

where

- M = Processability constant

3.3 Performance of Polymeric Composites

The advantages of composites are the ease of manufacturing, fabrication, and handling. Composites can be formulated and designed for high-performance applications along with durability. They have excellent specific mechanical properties and can be economically justified using the life cycle cost method. Some of the disadvantages of composites are high initial cost, creep, and shrinkage [11, 12]. The manufacture and fabrication require highly trained specialists from many engineering and material science disciplines. The composites are susceptible to environmental degradation, for example, alkalis and ultraviolet radiation damage polymeric composites. There are very little or nonexistent design guidance and/or standards for polymeric composites. They are difficult to join or adhesively bond.

The ability to tailor composites in the direction of a load or other operational parameter makes them the best-suited materials for engineering applications. Therefore, polymeric composite may be designed for many performances in a single component owing to their discrete properties and anisotropic behaviour. Lightweight airframes for aircrafts, damage-tolerant lightweight parts for automobiles, etc. have been made possible with advanced composites. Similarly, high-temperature airborne structures (such as ablative composites) are used in rocket motor linings, re-entry surfaces of rockets, etc.

Usually, matrix shrinkage is one of the major factors in the performance of polymeric composites. The matrix shrinkage causes inbuilt stress in polymeric composites. Therefore, the selection of the matrix, resin, and processing methods is the driving factor for the assessment and prediction of the performance of polymeric composites. Even so, there are many methods to monitor the condition of the polymeric composites. One of them uses different coloured laminae in the composite structure so as the composite weathers the new and emerging colour of ply indicates the remaining strength. During static or impact loading, the polymeric composite absorbs the energy by two methods: the emergence of new surfaces and material deformation. The first phenomenon that occurs is the deformation of the surface, and on the application of excessive stresses, a crack is initiated that creates new surface [13]. The propagation of cracks causes failure of the composite [14, 15]. While predicting the behaviour of polymeric composites, the following factors are important:

(a) Reduction in the bond strength of the interphase
(b) Loss of strength of reinforcement due to stress corrosion
(c) Loss of matrix strength
(d) Dependence of reinforcement and matrix on temperature, humidity, and time
(e) Degradation of the reinforcement and matrix from UV radiation

To Be Remembered

1. Micromechanics is the study of the behaviour of fibres and filaments, matrices, interfaces, and interphases in a composite upon the application of stress and strain.
2. Macromechanics is the study of the behaviour of the lamina or laminate when stresses or strains are applied.
3. Properties of composites are the sum of the product of volume fraction of the constituents and their properties.
4. Ideal volume fractions of reinforcement and matrix are 0.5.
5. In composites, a fibre-to-fibre load is transferred through the matrix.
6. Matrix experiences shear force during the load transfer from fibre to fibre in a composite.
7. Critical volume fraction of fibre is the minimum volume fraction of the fibre experiencing fibre-to-fibre shear force through the matrix and having good strength in the composite.

3

8. Shear force is at the maximum at the ends of fibres and is zero at the centre of the fibre in short fibre composites.
9. Tensile force is experienced at the maximum at the centre of the fibre and is zero at the ends of fibres in short fibre composites.
10. Composite laminae are orthotropic in nature (i.e. has a symmetrical structure in three principal planes).

Exercises

1. Calculate the density and tensile strength of E-glass-reinforced epoxy composite having V_f and V_m of 0.50 each. The density and tensile strength of E-glass fibres are 2.54 g/cm^3 and 3500 MPa, respectively. The density and tensile strength of the epoxy resin are 1.2 g/cm^3 and 80 MPa, respectively. What is the specific tensile strength of this composite?
2. How many layers of glass fibre are required for making a planar laminate of 3 mm thickness with ideal volume fractions? The thickness and arial density of glass fibre are 0.33 mm and 425 GSM (g/m^2), respectively. Densities of glass and epoxy resin are 2.54 g/cm^3 and 1.2 g/cm^3, respectively.
3. Calculate the random modulus of an injection-moulded glass fibre-reinforced nylon composite. Here, $V_f = 0.3$, fibre length = 2 mm, fibre diameter = 10 microns, $E_f = 350$ GPa, and $E_m = 3$ GPa.
4. Calculate the longitudinal and transverse modulus in GPa for a continuous fibre-reinforced composite having a fibre content of 60% by weight, $\rho_m = 1.0$ g/cm^3, $\rho_f = 2.5$ g/cm^3, $E_m = 3$ GPa, and $E_f = 70$ GPa.
5. A glass CSM-reinforced polyester composite has a 40% glass fibre by weight. The fibre length and diameter of glass CSM are 40 mm and 12 microns, respectively. Find the random modulus of this composite. The moduli of glass fibre and polyester resin are 72 GPa and 2.0 GPa, respectively.
6. Find the critical length of the fibre in a glass CSM-reinforced polyester composite. The ultimate tensile strength and diameter of the glass fibre are 3000 MPa and 12 microns, respectively. The yield stress of the matrix is 48 MPa.
7. What are the weaves or pattern of reinforcement to be selected for fabrication of (a) composite boat hull, (b) composite nose cone of aircraft, (c) composite water storage tank, and (d) composite pole/mast? Give reasons for your selection.
8. Water absorption of epoxy resin is 4%, and its specific gravity is 1.2 g/cm^3. What is the maximum water percentage in a carbon fibre-reinforced epoxy composite that has a specific gravity of 1.6 g/cm^3 and where $V_f = 0.5$.
9. Estimate the tensile modulus of a UD glass/epoxy composite and carbon/ aluminium composite for $V_f = 0.4$. Also calculate V_{crit} and V_{min} for the maximum load carried by these composites for a cross-sectional area of 20 cm^2.

Material	σ (MPa)	E (GPa)	ρ (g/cm^3)
Epoxy	35	4	1.2
Glass fibre	3000	70	2.5
Carbon fibre	3500	350	1.8
Aluminium	250	70	2.7

10. What are the failure modes in composites in tension and compression?

References

1. Agarwal, B.D. and Broutman, L.J., *Analysis and Performance of Fibre Composites*, John Wiley & Sons, USA, 1980.
2. Nielsen, L.E.; *Mechanical Properties of Polymers and Composites*, Vol-2, Ed., Marcel Dekker: New York, 1974.
3. Love, E.A.H., *A Treatise on Mathematical Theory of Elasticity*, Cambridge University Press, Cambridge, UK, 1927.
4. Varma, I.K. and Gupta, V.B., *The Science and Technology of Fiberous Composites*, Asian Books Pvt Ltd, New Delhi, India, 2014.
5. Shim, H.B., Seo, M.K., M., and Park, S.J., "Thermal conductivity and mechanical properties of various cross-section types carbon fiber-reinforced composites," *Journal of Materials Science*, 37 (2002), p. 1881.
6. Hasselman, D.P.H., Donaladson, H.K.Y., and Thomas, J.R.J., "Effective thermal conductivity of uniaxial composite with cylindrically orthotropic carbon fibers and interfacial thermal barrier," *Journal of Composite Materials*, 27 (1993), p. 637.
7. Halpin, J.C. and Tsai, S.W., "Environmental factors in Composite Design," *Air Force Materials Laboratory Technical Report*, AFML-TR-67-423, 1967.
8. Jones, R.M., *Mechanics of Composite Materials*, Mc Graw Hills, New York, 1975.
9. Spencer, A., "The transverse moduli of fiber composite material," *Comp. Sci. & Tech.*, 27 (1986), p. 93.
10. Chou, T.W., Flexible Composites, J. Mater. Sci., 24, 1989, 761.
11. Chou, T.W. and Kelly, A., "Mechanical Properties of Composites," *Ann. Rev Mater. Sci.*, 10 (1980), p. 229.
12. Gordon, J.E., *Structures*, Penguin: Harmondsworth, Middlesex, 1978.
13. Dwivedi, M., *High Performance Polymeric Nanocomposites: Radar Transparent Applications*, Lambert Academic Publishing, Germany, 2012.
14. Hull, D. and Clyne, T.W., *An Introduction to Composite Materials*, 2nd ed., Cambridge Solid State Science Series, Cambridge University Press, Cambridge, 1996.
15. Nye, J.F., *Physical Properties of Crystals: Their Representation by Tensors and Matrices*, Clarendon, Oxford, 1985.

Rheology in Processing of Polymeric Composites

A. K. Ghosh, M. Dwivedi, *Processability of Polymeric Composites*,
https://doi.org/10.1007/978-81-322-3933-8_4

Processing polymer composites is quite ubiquitous in the domain of polymer engineers. The true aspects of polymer processing have a direct correlation with rheological properties of the material concerned. The fundamental concept behind the theory of rheology dates back to the basic definition of work being done, which is dependent on two important parameters, notably *force* and its corresponding *displacement*.

$$W = F \cdot d$$

The force that is needed to displace a material from its coordinates and its centre of gravity to a new set of coordinates needs to overcome a *resistance* against the material to move from its position of inertia. Relating to Newton's first law of motion, this resistance is the principal parameter that governs this entire theory. The work done corresponds to the energy that is needed to displace the body, which also goes hand in hand with the parameter known as resistance.

This concept has been developed over time in order to define the intrinsic properties of a material, known as *rheology*, defined as the study of the flow and deformation of a material. The *resistance* talked about earlier corresponds to what is needed to make a material flow (for melts) or deform (for solids). Correspondingly, the work done can be attributed to the power required to accomplish the desired work. In simpler words, the significance of rheology in polymer processing is mainly to make a product easily with less power required [1].

The control of viscosities, kinetics of impregnation, and consolidation of a composite optimize both the processing time and the quality of composite products. In the processing of composites, the following phenomena occur:

- Capillary effects and dynamic wetting phenomena
- Fluid flow in a porous medium
- Mechanical equilibrium between the reinforcement and the fluid flowing through it
- Occurrence of mass/heat transfer mechanisms arising from curing or solidification

All of these phenomenon and mechanisms influence the rheology of resin during the processing and post curing of composites. Inapt mass/heat transfer, during curing, causes warpage in the composite product. Liquid resins and molten polymers are non-Newtonian fluids; therefore, only non-Newtonian flow takes place in the processing of polymeric composites [2, 3].

4.1 Fundamentals of Rheology

The study of flow or/and deformation of material (fluid/solid) is called rheology. Viscosity is part of rheology. *Viscosity* is defined as the resistance to flow of a fluid. Flow can be turbulent or laminar.

The *Reynolds number* of a fluid differentiates between turbulent and laminar flows.

$$\text{Reynolds number} = \rho v d / \eta \tag{4.1}$$

where
- ρ = Density
- v = Velocity
- d = Diameter of flow
- η = Viscosity

If the value of the Reynolds number is ≥2100, the flow is called turbulent; otherwise, the flow is called laminar. Polymers, either in the form of viscous resin or molten polymer, have laminar flow only [4]. Usually, the Reynolds number for liquid resins and molten polymers is <100.

4.1.1 Viscometry

4.1.1.1 Newtonian and Non-Newtonian Behaviour

Newton's law of fluids states that shear stress is directly proportional to the shear rate and the proportionality constant is called viscosity (η)

Shear stress and shear rate

$$\begin{aligned} \tau &\propto \dot{\gamma}\left(= dv / dt\right) \\ \tau &= \eta \dot{\gamma} \end{aligned} \tag{4.2}$$

where
- τ = Shear stress
- $\dot{\gamma}$ = Shear rate
- η = Viscosity coefficient

The viscosity coefficient, η, is known as the viscosity of a fluid. In *Newtonian* fluids, the shear rate of a fluid under isothermal conditions remains directly proportional to the shear stress at any point of consideration, as shown in ▫ Fig. 4.1. In *non-Newtonian* fluids, the shear stress is not directly proportional to the applied shear rate such that, when the shear stress is plotted against the shear rate, the points are not in a straight line. The Weissenberg effect is seen in non-Newtonian fluids. In this effect, a non-Newtonian fluid climbs up the stirrer rod when it is stirred.

Point to Ponder...

Tomato sauce shows thixotropic behaviour. It does not come out of the bottle instantaneously. Rather, with time its viscosity comes down slowly when it starts flowing out of the bottle.

4.1.1.2 Shear Flows and Deformation

Non-Newtonian fluids have different categories as discussed here.

Viscoelastic Fluids Viscoelastic fluids are those which behave as viscous liquids and elastic solids simultaneously and also exhibit partial elastic recovery after deformation.

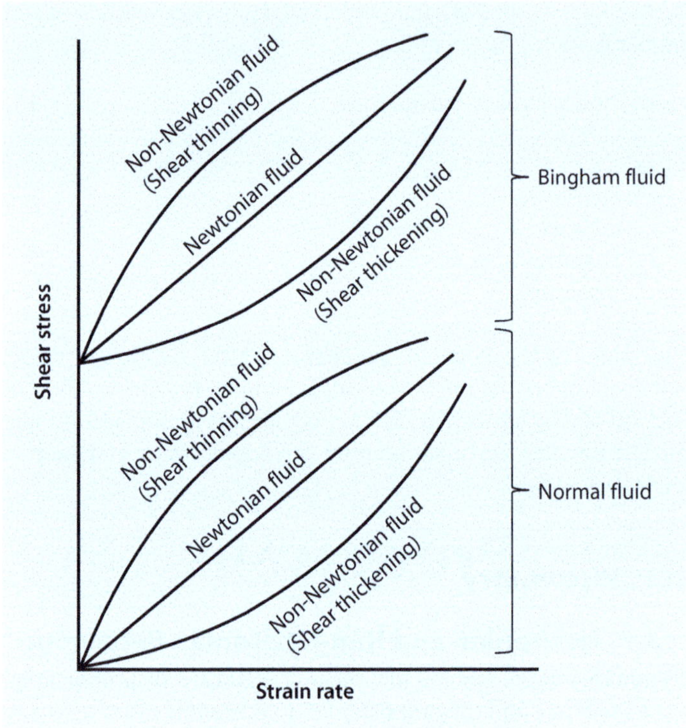

Fig. 4.1 Shear stress and strain rate relations of Newtonian and non-Newtonian fluids

Time-Dependent Fluids For these fluids, the apparent viscosity does not only depend on the shear rate but also on time the shear stress is applied. Time-dependent fluids may be divided into two distinct types.

- *Thixotropic Fluids*—In these fluids, shear stress decreases with time when the fluid is sheared at a constant rate. The structural breakdown in these fluids is a function of the duration of shear as well as shear rate. In thixotropic fluids, structures along with secondary bonds and intermolecular attractions would be progressively broken down, leading to the reduction of apparent viscosity. For most thixotropic fluids, the viscosity decreases with higher rates of shearing but only after the threshold level of shear occurs. Mechanical shearing alters the arrangement of the long chain of the molecular structure of these fluids. Mayonnaise and yogurt are examples of thixotropic fluids where shearing causes the attraction of the long chain of the molecular structure.
- *Rheopectic Fluids*—These are the fluids in which structure builds up from shearing with time. This phenomenon is the reverse of thixotropy. In rheopectic fluids, viscosity increases with time.

Point to Ponder...

Paints are Bingham plastic fluids. We have experienced that behaviour when brushes soaked in paint move on the wall or furniture after exerting initial necessary stress to overcome the inertia.

Time-Independent Fluids For these fluids, the shear rate is only a function of shear stress. Time-dependent fluids may be divided into three distinct types:

- *Bingham Plastic Fluids*—These fluids exhibit a yield stress that must be exceeded before flow starts. Tooth paste, oil paints, etc. show such behaviour.
- *Pseudoplastic Fluids*—The viscosity of pseudoplastic fluids decreases with an increase in shear rate. Molten polymers have this characteristic. Pseudoplastic fluids are used to advantage in injection moulding as the viscosity of the melt further decreases when the material flows through small cross-sectional gates.
- *Dilatant Fluids*—In these fluids, the apparent viscosity increases with an increase in the shear rate. Dilatant fluids are also referred to as *shear-thickening fluids* (STF). Cornstarch mixed in water is an example of a dilatant fluid.

Point to Ponder…
Shear-thickening fluids have a potential use in protective armour against stabs and bullets due to their property of thickening with increased shear rate.

4.1.1.3 Viscometric Models

The following viscometric models for different type of fluids have been developed:

- *Power Law Model*—This model requires two parameters to characterize any polymer. The power law model was developed by Ostwald-de-Waele and is applied for shear-thinning and shear-thickening fluids only. The equation governing this model is

  $$\tau = m\dot{\gamma}^n \tag{4.3}$$

  where
  - m = Consistency index
  - n = Power law index
- If $n = 1$, then $m = \eta$, then it is Newtonian fluid
- If $n > 1$, then it is shear-thickening fluid
- If $n < 1$, then it is shear-thinning fluid
- Eighty to eighty-five percent of polymers have shear-thinning behaviour upon melting, and the value of n for them varies from 0.4 to 0.7.
- *Truncated Power Law*—This model requires three parameters to characterize any polymer. This law is written as

  $$\eta = \eta_0 \tag{4.4}$$

  for $\dot{\gamma} \le \dot{\gamma}_0$

  $$\eta = \eta_0 \left(\frac{\dot{\gamma}}{\dot{\gamma}_0} \right)^{n-1} \tag{4.5}$$

  for $\dot{\gamma} \ge \dot{\gamma}_0$

  where
  - η = Viscosity
  - η_0 = Initial viscosity
  - $\dot{\gamma}$ = Shear rate
  - $\dot{\gamma}_0$ = Initial shear rate

The three parameters are η_0, $\dot{\gamma}_0$, and n.

- *Bird–Carreau Model*—This model overcomes the limitations of independence of both temperature and molecular weight distribution of the power law and the truncated power law. The Bird–Carreau model is dependent on temperature and molecular weight distribution. This model has four parameters and may be written as

4

$$\frac{\eta_0 - \eta_\infty}{\eta_0 - \eta_\infty} = \left[1 + (\lambda \dot{\gamma})^2\right]^{(n-1)/2} \quad (4.6)$$

where

- λ = Time constant
- η = Viscosity
- η_0 = Initial viscosity
- η_∞ = Infinite viscosity

The *time constant*, λ, is the time after which shear thinning starts and any change in molecular configuration becomes effective. The melt flow index is the indication for molecular weight.

- *Ellis Model*—This model has three parameters and may be written as

$$\frac{\eta_0}{\eta} = 1 + (\tau / \tau_{1/2})^{(n-1)/2} = 1 + (\tau / \tau_{1/2})^{(\alpha-1)} \quad (4.7)$$

where

- $\alpha = 1/n$
- τ = Shear stress
- $\tau_{1/2}$ = Shear stress at $\eta = \eta_0/2$
- λ = Time constant
- η = Viscosity
- η_0 = Initial viscosity

4.1.1.4 Factors Affecting Viscosity

The factors which affect the viscosity of composite medium are as follows:

- *Temperature*—Temperature causes the separation of polymeric chains when thermal energy surpasses the bond (secondary bonds) energy among the polymeric chains. Therefore, the viscosity of the composite mix at high temperatures is reduced.
- *Molecular Weight*—Higher molecular weight signifies longer polymeric chains; hence, more sites are available for intermolecular attraction on the polymeric chain of a polymer. This causes high-melt viscosities or high temperatures to overcome the intermolecular attraction among the polymeric chains of the polymer.
- *Pressure*—High pressures lead to high viscosities of polymer melt or liquid resin because the pressure puts polymeric chains under a physical constraint to move freely. This leads to an increase in the viscosity of liquid resin and polymeric melt under pressure.

- *Additives*—Additives in the form of fillers increase the viscosity of polymer melt. Additives in the form of plasticizers decrease the viscosity by breaking the long molecular chains of the polymer, thus reducing the molecular weight too.
- *Reinforcement*—Reinforcement causes a physical barrier; therefore, they increase the viscosity of the composite mix.

4.1.1.5 Normal Stresses

In non-Newtonian fluids, some radial outside force is exerted by the fluid in the pipe through which flow is taking place. This force is normal to the flow and is considered a normal stress exerted on the fluid or body. For the purpose of illustration, a surface ABCD is used, and this surface is in the *yz* plane perpendicular to the *x*-axis. τ_{xx}, τ_{yy}, and τ_{zz} are the normal stresses working on the surface or body, as shown in Fig. 4.2.

If total pressure acting on the body is *p*, the following equations for normal stresses may be written as

$$\tau_{xx} = -p + \tau_{xx} \tag{4.8}$$

$$\tau_{yy} = -p + \tau_{yy} \tag{4.9}$$

$$\tau_{zz} = -p + \tau_{zz} \tag{4.10}$$

Normal stresses τ_{xx}, τ_{yy}, and τ_{zz} act away from the body. Two normal stresses are denoted as N_1 and N_2. N_1 is the *first normal stress difference* (FNSD), and N_2 is *second normal stress difference* (SNSD), so

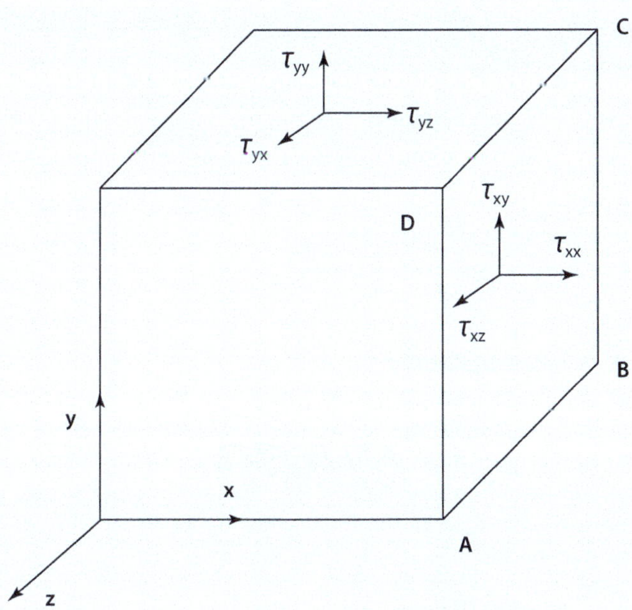

Fig. 4.2 Normal stresses working on a body

$$N_1 = \tau_{11} - \tau_{22} \tag{4.11}$$

$$N_2 = \tau_{22} - \tau_{33} \tag{4.12}$$

where 1 is the primary direction, 2 is the secondary direction, and 3 is the neutral direction. Two new rheological properties may be derived based on the previous equations. They are the FNSD coefficient and the SNSD coefficient:

$$\Psi_1 = \text{FNSD coefficient} = N_1 / \dot{\gamma}^2 \tag{4.13}$$

$$\Psi_2 = \text{SNSD coefficient} = N_2 / \dot{\gamma}^2 \tag{4.14}$$

The *relaxation time*, λ, is defined as a ratio of viscosity to the rigidity modulus:

$$\lambda = \frac{\eta}{G} \tag{4.15}$$

The rigidity modulus is given as

$$\text{Rigidity modulus}(G) = \frac{r^2}{N_1} \tag{4.16}$$

Therefore, the relaxation time may be given as

$$\lambda = \frac{N_1}{\tau\gamma} \tag{4.17}$$

The relaxation time, λ, is useful to find out the time required for a material to cool so that stresses—especially residual stresses after moulding—are relieved. The Deborah number is ratio of relaxation time and process time, so

$$N_{\text{Deb}} = \text{Deborah Number} = \frac{t_{\text{f}}}{t_{\text{p}}} = \frac{\text{Relaxation time}}{\text{Process time}} \tag{4.18}$$

where

- t_{f} = Characteristic fluid time
- t_{p} = Characteristic process time

4.1.2 Viscoelasticity

Viscoelasticity is made out of two words: viscosity and elasticity. It is the property of materials that exhibit both viscous and elastic characteristics when undergoing deformation. In other words, the viscous part of a material signifies its flow under stress, whereas the elastic part signifies the recovery of shape after stress is removed.

4.1.2.1 Model Flows

In rheology, there are two main driving forces that cause flow to occur. These forces are pressure difference and drag force.

- *Pressure Difference*—This could be understood with a model flow through a circular pipe as

$$\Delta P = P_1 - P_2 \tag{4.19}$$

where P_1 is high pressure and P_2 is low pressure.

The pressure gradient is given as

$$\text{Pressure gradient} = \frac{\Delta P}{l} \tag{4.20}$$

where l is the length of the pipe.

The velocity profile of a fluid flowing through a pipe may be parabolic, a straight line, concave headed, etc., depending upon the polymer. If water is flowing through a pipe, the velocity profile is parabolic. For the flow in pipe, the volumetric or mass flow rate, Q, is

$$Q = \int_0^R (\text{Local area} \times \text{Local velocity}) \tag{4.21}$$

where R is the radius of the pipe.

The cylindrical coordinate system (r, θ, z) shown in Fig. 4.3 is used for determining the boundary condition for flow through a pipe. Here R is the radius of the pipe, r is the radius at any instance in the flow, θ is circumferential portion, v_z is the velocity in the z direction, and z is the direction of the flow. The velocity of fluid in the z direction is depicted by v_z. But v_z is varying from the centre to the wall of the pipe. It also varies with respect to the direction i.e., $v_{z(r)}$, so

$$V_z(R) = 0 \tag{4.22}$$

if $r = R$ (velocity at wall of the pipe).

$$V_z(0) = V_{max} \tag{4.23}$$

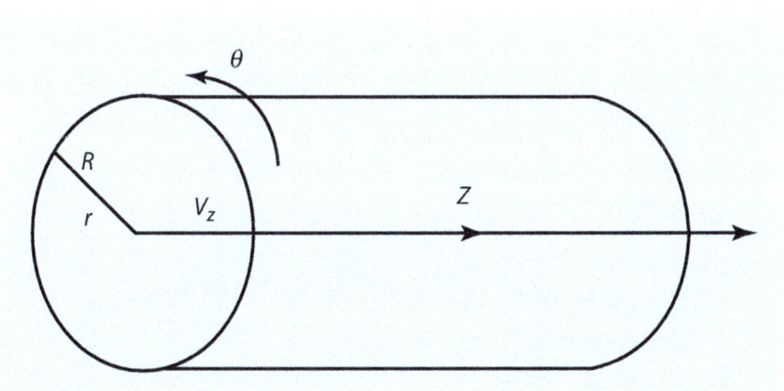

Fig. 4.3 Flow through a pipe

4

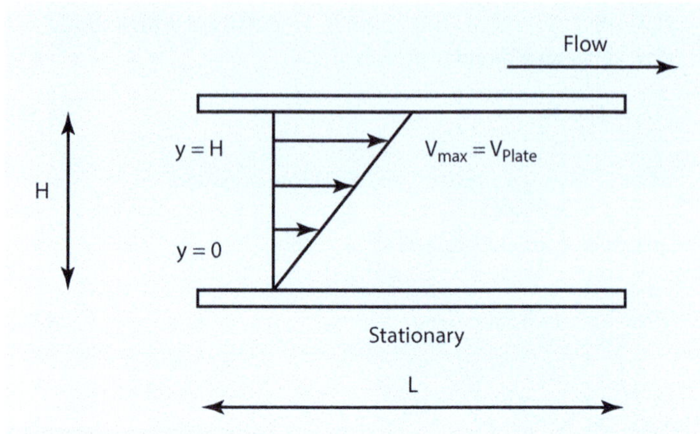

Fig. 4.4 Flow between parallel plates

if $r = 0$ (velocity at the centre).

The velocity profile is given as

$$V_z = V_{max}\left[1-\left(r/R\right)^2\right] \tag{4.24}$$

$$V_z = 0 \text{ at } r = R$$

$$V_z = V_{max} \text{ at } r = 0$$

Then,

$$V_z\left(\text{at}\, r = R/2\right) = 3/4V_{max} \tag{4.25}$$

- *Drag Force*—The other driving force in rheology is the drag force (e.g. shear driven flow) that could be well understood by the model of flow between parallel plates using a linear coordinate system (x, y, z). The flow between parallel plates is shown schematically in Fig. 4.4. The width and thickness of the plate are w and t, respectively.

The velocity profile for flow between parallel plates may be formulated as

$$V_z = V_{max}\left(y/H\right) \tag{4.26}$$

The volumetric flow rate for flow through a circular pipe using Eq. 4.24 is given by

$$Q = \frac{\pi R^4\left(\Delta P\right)}{32\eta L} \tag{4.27}$$

where
- Q = Volumetric flow rate
- R = Radius of pipe
- ΔP = Pressure difference
- L = Length of Pipe
- η = Viscosity of fluid

Then,

$$\eta = \frac{\pi R^4 (\Delta P)}{32 Q l} \tag{4.28}$$

Equation 4.28 can be rearranged as

$$\eta = \frac{\frac{R}{2}\left(\frac{\Delta P}{L}\right)}{\frac{32 Q}{\pi R^3}} \tag{4.29}$$

Using force balance in a circular pipe, it can be obtained that shear stress at the wall is given b,

$$\tau\left(\text{at wall of tube}\right) = \frac{R}{2}\left(\frac{\Delta P}{L}\right) \tag{4.30}$$

Hence, the shear rate is

$$\dot{\gamma} = \frac{32 Q}{\pi R^3} \tag{4.31}$$

As the fluid considered for this analysis is Newtonian, the shear rate and viscosity are called the *apparent shear rate* and *apparent viscosity.*

4.1.2.2 Shear Free Flow

It has been discussed previously that the rheological properties of polymeric matrix in composites have a direct impact on many manufacturing processes. Shear free flows are also known as elongational flows or stretching. They are a specific characteristic of polymeric matrices, and polymer melts due to their viscoelastic behaviour. This is the flow in which it is possible to select elongational stresses for every fluid element and orthogonal set of unit vectors (shear stress) so that the strain rate also has diagonal components. These flows have no non-zero, off-diagonal components in shear stress (τ), and the shear rate (γ) in elongational flow is shown as

$$\begin{pmatrix} \tau_{11} & 0 & 0 \\ 0 & \tau_{22} & 0 \\ 0 & 0 & \tau_{33} \end{pmatrix} . \begin{pmatrix} \gamma_{11} & 0 & 0 \\ 0 & \gamma_{22} & 0 \\ 0 & 0 & \gamma_{33} \end{pmatrix} \tag{4.32}$$

In elongation flows of incompressible fluids such as polymer matrix and polymer melts, only ($\tau_{33} - \tau_{11}$) and ($\tau_{22} - \tau_{11}$) can be measured. The elongational flow is shown schematically in ◘ Fig. 4.5. The flows occurring during these processes are complicated and will usually involve some combination of planar Couette flow (PCF), planar elongational flow (PEF), and biaxial/uniaxial stretching flow (BSF/USF). PCF is the laminar flow in the space between

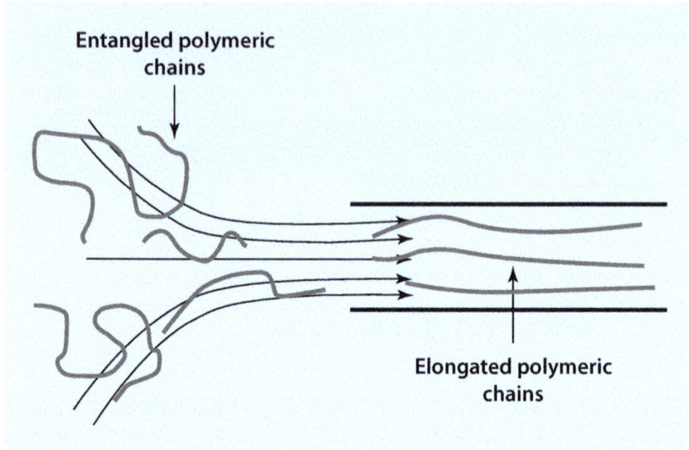

Fig. 4.5 Elongational flow

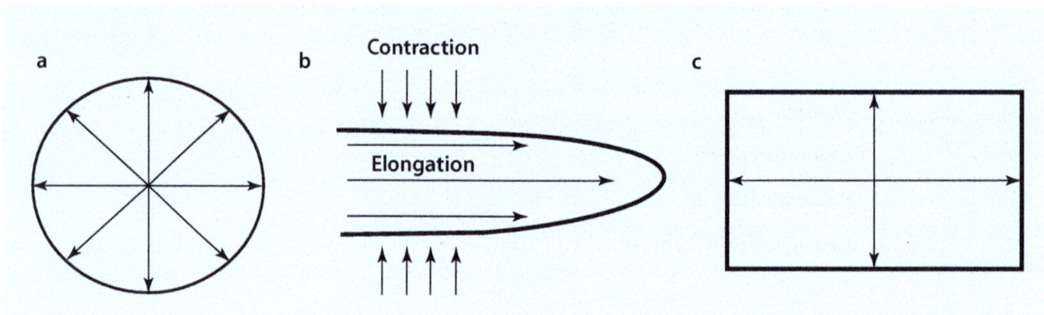

Fig. 4.6 Schematic presentation of **a** PEF, **b** USF, and **c** BSF

two parallel plates—one of which is moving relative to the other. The flow is driven by virtue of a viscous drag force acting on the fluid and the applied pressure gradient parallel to the plates [5]. The extension of the polymer occurs in all three directions in PEF, as shown in Fig. 4.6a.

Strong stretching occurs in one direction, while contraction occurs equally in the other two directions in USF, as shown in Fig. 4.6b. In BSF, the stretching of the polymer occurs in two directions, as in the cases of the blown film shown in Fig. 4.6c.

Normal stresses occur due to the nature of the fluid, but shear flow occurs due to the flow of fluid.

4.2 Rheology Aspects of a Polymeric Matrix

Since thermoplastics tend to have high molecular weights as well as a greater molecular weight distribution, it is evident that the nature of the polymer will be different from Newtonian behaviour. The polymer chains are inherently in an entangled state, which is by far the inertial state of the system. Application of shear at defined rates leads to disentanglement of the polymer melt.

Substances like water, which are composed of single unit, are essentially Newtonian as there are no entanglements present for the applied shear rate to make them disentangled. Such materials have constant viscosity across the entire window of shear rates. A drop in viscosity with an increase in shear rate, which is a symptom of disentangling of the entanglements, is called *shear-thinning behaviour*. As far as the melt rheology of polymer is concerned, the polymer shall start to show a shear-thinning behaviour after a critical shear rate. The resistance to flow (i.e. melt viscosity) of the polymer will drop eventually on the application of higher shear rates (analogous to applied force). This can have severe implications on the final product. The sole aim of processing thermoplastics is to make a moulded product. Owing to its shear-thinning nature, the viscosity of the polymer is much lower than the supposed viscosity—had the polymer been Newtonian. Eventually, lower viscosity leads to higher flow and, ultimately, higher output rate or mouldability.

Therefore, there is always a prior need to having the knowledge of the output rate of the melt from any polymer processing device. Since the output has a direct correlation to the velocity of the polymer melt when flowing and the velocity of the flowing polymer is directly dependent on the viscosity of the polymer at the processing conditions, the importance of rheology in governing the polymer processing methods is immense. Rheological analysis of thermoplastics can give a wide comparative knowledge about the processability of a particular polymer.

In order to develop a high-quality composite structure, there should be good bond between the fibres and the matrix. This requires proper wetting of the fibres by the matrix. Therefore, the matrix resin must flow. In the processing of composites, following four types of flows occur.

1. Resin percolation through and along the fibre
2. Transverse flow through the fibres (squeezing flow consisting of resin shear within the ply in the transverse direction)
3. Intraply shear along the fibre direction
4. Interply slip (cooperative flow among the layers with the layers in either the same or different orientations)

In polymeric composites, the *apparent viscosity* ($\eta*$) increases with an increase in fibre loading and decreases with an increase in temperature and shearing. This is due to the shear-thinning behaviour of the matrix. A greater amount of fibre loading may cause enhanced friction at the interface of the polymer and its fibres, which may hinder the chain mobility, gradually increasing viscosity. The incorporation of fibres in polymer systems generally increases the viscosity. The more nonpolymeric medium in the matrix, the more the hindrance for the movement of polymeric chains. The orientation of fibres strongly depends on the amount of fibre loading. At optimal or lower fibre loading, the flow of the polymer is more feasible and enables the fibres to align in the direction of flow. This in turn causes anisotropy in short fibre composites.

4.3 Rheology in Material Selection

One of the renowned rheological instruments, the melt flow rate (MFR) analyser is commonly used in plastic processing industries to have a prior idea about the viscosity of the melt. For example, if we have two thermoplastic materials (say polypropylene) with varying PDI (say 3 and 10) but with similar MFR value, we ought to have different eases of processability for the two polymers at a particular condition that eventually can be predicted by rheology. On comparing the viscometric plots of the two polymers, the zero shear viscosity values for the two polymers shall be very close. However, the critical shear rate beyond which the shear-thinning regime begins and the power law index, n, will vary significantly. The polymer with a higher PDI will have a critical shear rate much lower in comparison to the polymer having a lower PDI. Also, if a polymer has a narrower PDI, it will show a steeper viscometric plot in comparison to the polymer having a broader PDI.

To answer the question on how to process polymer composites efficiently, the understanding of rheology for the matrix phase is of utmost importance. This is mainly because the secondary phase and the fillers are going to get incorporated into the matrix. Commonly used matrices include polyethylene, polystyrene, etc. Polymers like polyethylene are known to have branching inherently in their polymer chain if no catalyst is used during synthesis. Whether the polyethylene will have branching or not will also govern the processability of the polymer composites. The presence of branching will increase the free volume of the macromolecules. In addition, it will contribute to the broadening of the molecular weight distribution thereby lowering die swell. The fact is that intermolecular vibration decreases in the case of branched polymers in comparison to linear polymers; therefore, the lowering of shear viscosity is observed in linear polymers for the same molecular weight. The lowering of melt viscosity leads to the ease of processability. This requirement of shear rate is lower than that for linear polymers of same molecular weight, and eventually it lowers the power consumption for a similar extent while mixing. The drop in shear viscosity of the polymer is in attendance only up to the point where the long chain branching does not contribute to the entanglement density of the polymer. However, if the lengths of the long chain branching increase beyond a critical limit, it may contribute in the entanglement density of the matrix. Hence, beyond a critical point in the lengths of long chain branching, the melt viscosity of the polymer shall start to show an increase.

Of the previously mentioned four flow patterns, resin percolation is the most important in achieving good bond between the fibres and the matrix. Darcy's one-dimensional law of flow through porous media for one-dimensional penetration is given as

$$X = \left(\frac{2kPt}{\mu} \right)^{1/2} \tag{4.33}$$

where
- x = Penetration depth
- k = Media permeability
- P = Pressure
- t = Time under pressure
- μ = Viscosity

This law shows that high viscosities will reduce the penetration depth unless the porosity, pressure, or time of the fibre bundle is increased accordingly.

The width of the molecular weight distribution also plays an important role in fibre wet out. If molecular weight distribution width is narrow, polymer will melt over a narrow range of temperature, and fibre wet out will be easier. This is seen in nylon matrix composites because nylon is highly crystalline thermoplastic; therefore, it shows sharp melting points which is a characteristic of crystalline material. Therefore, crystallinity also affects the ease of wetting out the fibre in composites.

A process for improving the ability of selected high-viscosity polymers to penetrate and coat fibres is based on the chemical technology known as 'cyclics'. In this technology, a high-viscosity thermoplastic polymer is initially formulated in a cyclical, short chain structure that has an inherent lower viscosity than the normal linear structure, and the molecular weight is simultaneously increased by heating the resin in the presence of an ionic catalyst. This ring-opening polymerization usually can be accomplished during the normal shaping in the moulding step.

4.3.1 Rheology in Deciding Processing Conditions

Rheology comes into the picture even at the compounding stage in an extruder where maintaining a temperature profile along the flow path is a stringent requirement. Processing polyolefins does not pose much difficulty due to their inherent ease of processability. Industries all around generally tend to maintain a linear temperature profile for them.

Materials that are inherently hard, such as polycarbonate, need to be processed in such a way that they do not affect the extruder. Hence, it is advisable to maintain a temperature profile with a small range that is much higher than their glass transition temperature (melting temperature for crystalline polymers).

A further advantage of a rheological study prior to processing is in the case of polymers like polyamides. Polyamides, commonly known as nylon, are blessed with an inherent structure of strong inter-chain hydrogen bonding, which is the main reason for their ultimately high mechanical properties. However, the polymer has a strange tendency to flow above its melting point. Therefore, it is very important to judge the temperature required to process nylon, as inadequate ideas of viscosity at various temperatures beyond its melting point may lead to a 'drooling' phenomena while processing, leading to an inefficient output rate of the resulting polymer. This calls for a requirement of a thorough viscometric plot analysis for polymers like nylon and judgement of its processing temperature prior to its processing. A corollary to this results in the selection of a temperature profile in the extruder that should be maintained in such a way to prevent the nylon from flowing like water, ultimately reducing the ease of processability.

Furthermore, the incorporation of fillers, such as glass fibre, that tend to break at high shear need to be carefully processed. Glass fibres are mostly fed into the extruder from a secondary feeder located somewhere amidst the flow path. The temperature across the primary feeder to the secondary feeder needs to be maintained in such a way that the glass fibres are fully wetted by the molten matrix. While incorporating inorganic particulate fillers from the primary feeder, it is of utmost importance that the temperature at the primary feeder junction is higher than the melting or glass transition temperature of the matrix. This way the filler gets a high enough percentage of the residence time to get impregnated by the molten polymer matrix.

4.3.2 Rheology of Incorporation of Elastomer

Rheology is also very important when it comes to incorporating elastomer into the matrix of thermoplastics. To disperse elastomer particles uniformly into the matrix of a thermoplastic, one has to implement high shear rates. High shear rates are responsible for developing an intermediate morphology that has elongated the dispersed elastomer particles throughout the polymer matrix. On application of a high shear rate for a sufficiently long time, the elongated elastomer particles stretch further in order to develop striations during the mixing stage. At the end of the mixing time, the striations get stretched to such an extent that they eventually collapse and yield elastomer particles of a fine size. The point to be noted is that the final dispersion and distribution of elastomer particles in the final polymer matrix is a strong function of shear rate, temperature, and a third factor, which is the difference in the viscosity of the thermoplastic as well as the elastomer. Elastomers are known to have much higher viscosities when compared to thermoplastics owing to their molecular weights. The higher the viscosity of the elastomer to be incorporated, the more difficult it is to get dispersed in the matrix. Hence a higher shear rate is needed that eventually calls for a higher power requirement. A thorough knowledge about the viscosity of the polymers at the processing temperature at high shear rates is necessary to ensure that a finely dispersed morphology is obtained at an optimum power consumption.

4.3.3 Rheology in the Dispersion of Fillers

Polymer composites are supersets comprised of polymers incorporated with fillers of varying sizes. Fillers with sizes in microns can give a totally different set of properties to the overall composites in comparison to fillers with sizes in the scale of nanometres. Today, composites based on fillers with sizes in the nanometres have been classified as a separate stream and are known as *nanocomposites*. The processing of composites with micron-sized fillers is known to be much easier than that of fillers in the nanometre scale.

Fibre-filled polymer melts have essential properties characteristic to them. Fibres are known to impart a mechanical integrity to the compound. Often, certain fibres may contribute to properties like conductivity owing to their nature and orientation within the polymer matrix. However, the role of fibres in the eventual rheological properties of the composite can be immense due to their higher L/D ratio. Higher L/D ratios sometimes cause problems with poor dispersion in the polymer matrix. This calls for an increase in the shear rate requirement for dispersing fibres within the polymer matrix. It has been found that sometimes—at a superior loading of fibres—the viscosity of fibre-based composites drops significantly at high shear rates.

4.4 Flow Behaviour When Processing Composites

The penetration of resin is more difficult in thermoplastics than in thermosets because the thermoplastics have inherently high molecular weights which lead to higher viscosities compared to thermosetting liquid resins. High viscosities of matrices cause difficulties in mixing and coating fibres. Whereas, thermosetting resins develop high molecular weight during the curing operation which is after the fibres have been impregnated/ coated by thermosetting prepolymer. Typical viscosities of matrices are given in ◘ Table 4.1.

Apart from the viscosity, the flow behaviour in processing composites depends upon the following parameters:

- Pressure difference applied
- Drag force assisting the flow

◘ **Table 4.1** Typical viscosities of matrices

Serial No.	Matrices	Viscosity (Pa.s)
1.	Thermosetting prepolymer (e.g. epoxy, phenol formaldehyde, polyester, polyimides, etc.)	0.5–15
2.	Thermoplastic polymer melt (e.g. polypropylene, polyethylene, nylon, polycarbonate, polyether ether ketone, polymethyl methacrylate, etc.)	1.5–30

4

4.4.1 Fibre Aspect Ratio

The *fibre aspect ratio* in the rheology of polymeric composites is important because it describes the behaviour of a material while undergoing processing as well as the sensitivity of the material to the microstructure and morphology. Although the length of the fibres is crucial for the reinforcing effect, the more important parameter is the fibre aspect ratio (length-to-diameter ratio). Usually, the fibres get aligned in the direction of flow. In the injection moulding processes, the alignment of fibres takes place along the wall of the mould due to shear force. Whereas in the core of the moulding, it is observed that the alignment of fibres is comparatively less when due to a reduction in shear force. The higher the fibre aspect ratio, the better the properties of composites. The most effective fibre reinforcement is achieved if the length of the fibre exceeds the critical fibre length, which is also dependent on the diameter of the fibre. Excessive shear force breaks the brittle fibres; therefore, the overall properties of the composite are reduced. In order to take advantage of the fibre length, the diameter of the fibre ideally should be decreased in order to increase the fibre aspect ratio, which governs the reinforcement efficiency of the fibres in the composites.

The use of rheological techniques enables a real-time monitoring of the processing influence on the microstructure of materials that has been extensively used in the study of composites in conjunction with basic characterization techniques (e.g. SEM, mechanical properties). The rheological methods enable the measurements of the properties of a polymer melt in response to both linear and nonlinear deformations.

4.4.2 Fibre Content

Due to fibre loading in composite mix, the viscosity of composite mix is expected to increase rapidly as the filler packs the composite mix more closely. Therefore, collisions among fibres increase, as discussed in ▶ Sect. 4.2. The *apparent viscosity*, η^*, increases at a low frequency because the fibre and polymer chains are not completely oriented during the dynamic frequency test. The addition of fibres in polymers perturbs the flow of the polymer and hinders the mobility of the polymer chain during the flow. With an increase in fibre loading, this phenomenon becomes more predominant, and the viscosity of polymer systems further increases. This observed behaviour is due to the formation of a strong network structure among fibres and polymeric chains. This causes resistance in flow because the rheological behaviour is sensitive to the molecular structure.

Continuous fibres or fabric offers very high resistance for the flow of matrices (i.e. melted thermoplastic as well as thermosetting prepolymers). The low viscosities of thermosetting prepolymers offer an advantage for the percolation of resin around the fibres in

Table 4.2 Generalized reinforcement content for thermosets and thermoplastics

Serial No.	Type of reinforcement	Matrix	Volume fraction (max)
1.	Fabric/continuous fibres	Thermoset	0.5
2.	Fabric/continuous fibres	Thermoplastic	0.5 by compression moulding or autoclaving
3.	Short staples (in the form of DMC/BMC)	Thermoset	0.5
4.	Short staples	Thermoplastic	0.3
5.	Short staples	Thermoset	0.4

fabric. This in turn assures the wetting of the fabric and results in a good composite after moulding. All high-performance and high-end applications in composites involve reinforcement in the form of fabric, and thermosetting resins are largely used for their fabrication. Compression moulding and autoclaving are the processes used for fabric-reinforced composites when external pressure is applied to fabricate the thermoplastic composite parts.

The surface treatments used on hydrophilic fibres make them hydrophobic; thereby, their wettability increases with hydrophobic matrices. If the fibre treatment is not properly applied, either a small fragment of fibres break out or scaling on the fibres comes out, which is called *fuzzing* in the case of carbon fibres. This creates resistance in the smooth flow of the matrix through microspaces among the fibres and causes an increase in the viscosity of the composite mix. The reinforcement content in a matrix is denoted in terms of the volume fraction of the composite. The volume fraction of reinforcement is 0.50 (in an idealistic situation) in order to achieve maximum properties in composites. The generalized reinforcement content for thermosets and thermoplastics are provided in Table 4.2.

Values of the volume fractions of reinforcement depend upon the viscosity of the matrix and processing method employed. In the processing of composites, application of high temperature and external pressure causes a reduction in viscosity. The weave of the fabric also contributes to its pliability, drapability, and wettability. Tightly woven fabrics (such as mat, basket, or twill weaves) offer resistance in the flow of the matrix, whereas satin weaves offer a smooth flow of the matrix. The twist in the yarn further restricts the flow of the resin and affects the processability of the composite. On the other hand, woven rovings of reinforcement have loose bundles of fibres and facilitate the smooth flow of resin among them.

4

4.4.3 Dependence of Processing on Rheology

The major constituents of polymeric composite are the reinforcement and the polymeric matrix. The role of reinforcement during processing is minimal as long as its physical orientation is maintained, and it is chemically and thermally inert. Usually, most of the reinforcements are thermally and chemically inert during the processing time and increase in temperature for polymeric composites. Of course, the particulate and short fibre reinforcements in the matrix affect the viscosity of the overall mixture, so the viscosity is increased. However, the increase in viscosity applies to the matrix only. Therefore, the major role in the processing of polymeric composites is played by the matrices because its physical state from liquid to solid determines the completion of the processing of the polymeric composite. The fabrication process is dependent upon the behaviour of polymers in terms of the response of flow with time and temperature.

The aim of any fabrication process is to achieve a high production rate with a reduction in processing time. Therefore, the curing or solidification of the matrix has to take place at the fastest rate without any defect occurring in the part. It is mandatory that temperature and time—both—need to be optimized to shorten the processing time. The flow of the matrix needs to be assured to percolate throughout the reinforcement to wet it. The lowest viscosity of the matrix resin enables easy percolation of it around the fibres to facilitate their wetting. Soon after the complete wettability is achieved, the solidification of the matrix resin must take place to form the desired composite.

It is amply evident from this discussion that the flow of the matrix in processing controls the overall quality of the composite, including the processing time and cost.

4.4.4 Change of Viscosity with Time and Temperature

Time and temperature both have a profound effect on the viscosity of the matrix in composites. Within the elapsed time, the viscosity of the composite system—cured or solidified—increases due to the advancement of cross-linking of the polymeric network in thermosetting matrices and also with solidification to the melt in thermoplastic matrices.

However, the effect of temperature is reversed in the case of thermoplastic matrices. With increasing temperature, the viscosity of thermoplastic matrices comes down. Also in the case of thermosetting matrices, the viscosity drops initially at moderate temperatures due to the physical separation of the polymeric chains of the matrix, which helps in the percolation of the matrix around the fibres in composites. However, by increasing the temperature further, the viscosity starts increasing due to the onset of cross-linking of the polymeric network structure. The viscosity behaviour in polymeric matrices is shown in ◘ Fig. 4.7. The temperatures

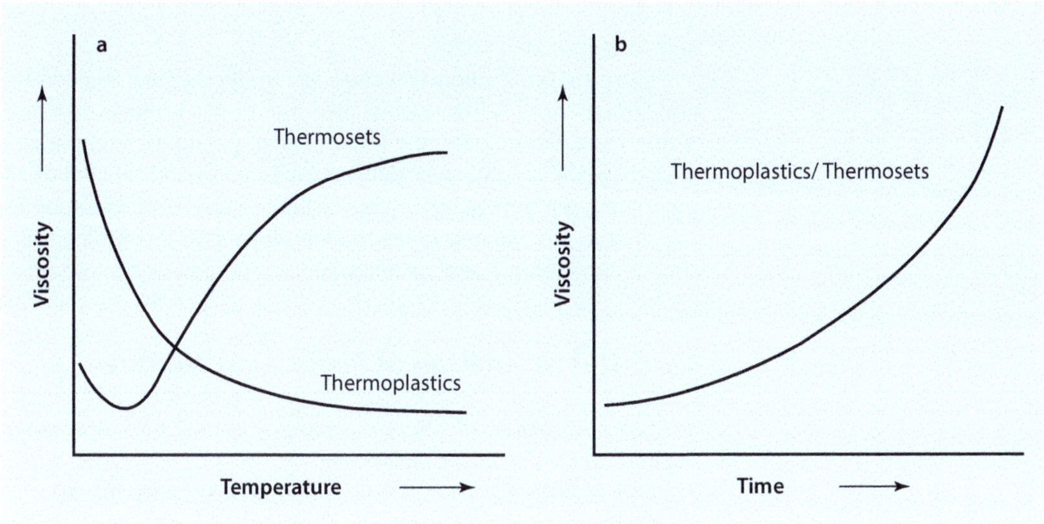

Fig. 4.7 **a** Viscosity behaviour of matrices with temperature and **b** viscosity behaviour of matrices with time

discussed here are within the limits of curing temperature, which are well below the thermal degradation temperatures. Thermosetting matrices in solid form are called *novolacs*. Novolacs initially melt like thermoplastics on the application of heat and then solidify into a polymeric cross-linked network structure of thermosets. Hence, the viscosity drops initially as they behave as thermoplastics, but with time and temperature, the viscosity increases infinitely due to cross-linking and solidification.

4.5 Rheology and Processability

It is clear from the previous discussion that the flow of matrices has an important bearing on the processing of polymeric composites. The rheology of the matrix or the composite mixture is greatly responsible for the overall properties of the resultant composites. This also decides the extent of the processability of polymeric composites.

4.5.1 Optimization of Process Conditions

Rheology of the matrix in combination with reinforcement controls the processing of composites. As a matter of fact, in the processing of polymeric composites, the matrix is initially brought to a low viscosity in order to wet the reinforcement and to solidify it to attain the structural integrity for the requisite physical, mechanical, thermal, and electrical properties. The optimization of the process conditions implies that the viscosity must not be brought down so much that the matrix settles down with gravity, leaving the resultant composite resin starved. Also, bringing the viscosity much too low with the application of heat may cause thermal degradation and increase the cost of processing by consuming more electricity. High consumption of electricity adversely affects the cost of production and environmental safety.

There are other important process conditions, such as pressure and temperature. A low viscous medium creates uniform pressure inside the mould during the processing of polymeric composites. Control of the mould temperature determines the rate of cooling of the matrix, and low temperatures facilitate a faster solidification of the matrix. The balance between solidification and complete wetting of the reinforcement or filling of the mould must be ensured.

4.5.2 Determination of Processing Window

The determination of the processing window for composites depends upon the following factors:

1. *Type of Reinforcement*—Choice of reinforcements influences the processability of polymeric composites. The volume fraction of the reinforcement and its surface treatment influence the overall flow of the resin in the polymeric composites. Short fibres/staples are suitable for injection moulding, whereas long fibres in any form and shape (such as fabrics or rovings/tows) are used in all other processing technologies. The material of fibres is another determining parameter to be used for a particular processing technology. Processing technologies that employ high temperatures use primarily carbon or graphite, glass, and aramid (Kevlar®) fibres. Of late, with the invention of new reinforcing materials (such as alumina whiskers, tungsten carbide, polycarbosilanes, etc.), the range of reinforcing material for high-temperature processing has increased. These materials have not been exploited for their full commercial potential due to their high costs, although they have good properties. Some fibres (such as polyethylene/UHMWPE or polypropylene fibres) are low melting temperature fibres and are not suitable for high-temperature processing technologies. Their processability is limited to moderate temperatures (i.e. < 120 °C). Natural fibres (such as jute, flex, bamboo, etc.) are not processed at high temperature either. Particulate reinforcements (such as carbon black, minerals, and most of the chemical compounds) are stable at high processing temperatures; therefore, these are used widely. In short fibre- or particulate-reinforced composites, the size of a suspended particle, size distribution, and volume fraction have influence over rheology concerning the processability of polymeric composites.
2. *Type of Resin System*—Resin systems affect the processability of polymeric composites in two ways: affinity to form bonds with fibre and by providing a desired shape of the product. All organic resins are hydrophobic in nature; therefore, they don't have affinity towards inorganic fibres that are hydrophilic in nature. The process of transforming the nature of inorganic fibre from hydrophilic into hydrophobic will be discussed in the next section.

High-temperature curing matrices adversely affect low melting temperature fibres. All of the organic matrices (barring polyolefins, which are inherently inert) form a good bond with commercially available reinforcing fibres. Polymeric matrices depend upon the polymer type, chain architecture, molecular weight, and distribution. The role of branching in polymeric chains has an influence on the viscosity of the resin. Highly branched resin systems have high viscosities due to the formation of large macromolecules of the polymer. The degree of orientations is determined by the rheological behaviour of the polymer and nature of the flow in the fabrication process. Molecular orientation hence influences the mechanical properties of moulded composite products.

3. *Type of Coupling Agent*—Inorganic fibres are treated with coupling agents or provided with surface treatments to make their surfaces hydrophobic. The hydrophobic surface of the fibre allows it to bond with a hydrophobic resin system. In the case of carbon and with the help of chemical processes, reactive groups (such as -OH, -COOH, -HCO, -CO, etc.) are attached. Once the surface of carbon fibre is reactive, a coating with a thickness of a few micron (2–3 μm) of either epoxy resin or polyimide is applied to make it hydrophobic. The surface of an aramid is inherently chemically active and hydrophobic, but it has very high smoothness; therefore, the surface energy is the least, which does not allow the matrix to adhere to it. Therefore, a surface treatment is used to roughen up the surface of aramid fibre with the help of an emery stone. The roughening of an aramid surface causes the undulation and surface defects that may lead to a weak composite. In view of this, another technique is used for making the surface of an aramid fibre bond to the resin. In this technique, an aramid fibre is oxidized by passing it through bromine water or a very dilute nitric acid. The oxidized surface of the aramid fibre has an even roughening in this process. In order to avoid deterioration of its surface further, a thin coating (2–3-μm-thick) of polyimide is applied over its roughened surface.

   In the case of glass fibres, the sizing is applied over the fibres. The sizing materials used are starch or polyvinyl alcohol. The role of the sizing material is to prevent the breakage of the glass fibres while handling or weaving. No sizing material is needed for surface treatment of carbon or aramid fibres owing to their strengths and moduli. Once glass fibres are brought into the desired form of fabric or rovings, the sizing is washed away, and the washed fabric is treated with silane or titanate coupling agents. These organometallic compounds are attached to the surface of glass from its inorganic part and leave open the organic part to make a bond with the organic resin system.

   Fibres such as polyethylene and polypropylene are highly inert, although they are organic in nature. The surfaces of these fibres are discharged with a corona to make them rough so that mechanical locking with the resin takes place. Hardly any chemical bonds are expected in their composite. At best, only some secondary bonds or van der Waals forces may occur.

4. *Product Design*—The rheology of matrix resin depends upon its flow path, which is governed by the product design. Not only is the processability of the composite product dependent on the product design but also the quality. It can be said if the product design is not conducive to the processing of the composite product, the rate of production and quality of the product are likely to suffer. The guiding factors for product design are the following:
   - Functionality
   - Service environment
   - Cost
   - Assembly
   - Material properties
   - Service life
   - Life cycle cost and maintenance
   - Transportability
   - Government regulations
   - Surface finish, aesthetics, etc.

   The interrelation of product design with raw materials, processing, and the final product is shown in Fig. 4.8.

   The product design has to have features that cause smooth flow of the resin with complete wetting of the reinforcement. The features of the product design and its effects on composite products are given in Table 4.3.
5. *Mould Design*—Product designs, raw materials, and processing technology dictate the mould design. These parameters are interrelated for product design as shown in Fig. 4.8. High viscosities are not favoured for the processing of polymeric composites because they don't allow efficient wetting of the

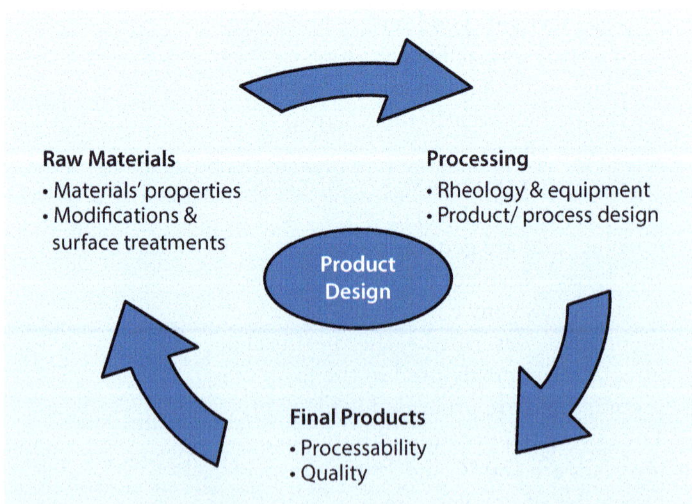

Fig. 4.8 Interrelation of composite product design with raw materials, processing, and the final product

Table 4.3 Product design and its effects on a composite product

Serial No.	Feature of product design	Effect on processability and performance of the end product
1.	Sharp corners	Region would remain resin-rich; hence, weak in mechanical properties
2.	Round surfaces or round corners	Enable the smooth flow of resin and necessary reinforcement is expected
3.	Thin section	Warping may occur. Requisite volume fraction of the reinforcement is a challenge
4.	Thick section	Good quality in laminated composites, but complete curing in thermosetting resin remains a challenge. In injection-moulded composites, the orientation of fibres is achieved in the direction of flow only on the surface, and orientation of fibres in the core is lacking
5.	Undercuts	Designs having undercuts are not preferred in composites. Undercuts complicate the mould design. Sliders are used in the mould design for unmoulding a composite product having an undercut. Often, a composite part is mechanically weak due to the low fibre volume fractions at these locations

fibres. The inefficient wetting is largely operative in a closed and confined mould. Therefore, in cases where a high viscous resin is used, the mould must have enough provisions for uniform heating for the reduction of viscosities to make the composite mix processable.

6. *Moulding Process*—The moulding process is the method to convert the raw materials into a composite product. This is the most important processing window as it determines the rate of production, cost of production, and quality. The processability of raw materials is intertwined with the processing method or moulding process. A particular form of raw material is processable with a particular processing technology, as given in Table 4.4.

Some moulding processes support the processing of highly viscous resin systems. Moulding processes (such as thermoforming and compression moulding) are open moulding processes. Short fibre-reinforced, semi-crystalline or amorphous thermoplastics may be moulded using these processes. UD fabric-reinforced thermoplastics are capable of being thermoformed in the normal direction of the fibre orientation.

Most of the moulding processes are closed moulding processes in a confined space, and they are used for low viscous resin systems during the processing. There is a myriad of choices of moulding processes for polymeric composites. These are discussed in detail in ▸ Chaps. 7 and 8.

Point to Ponder...
Incompatibility of an oil and water mixture is due to their hydrophobic and hydrophilic natures, respectively.

4

Table 4.4 Processability of raw materials and corresponding processing methods for polymeric composites

Serial No.	Processing method	Form of raw material
1.	Filament winding	Reinforcement: filaments Matrix: liquid
2.	Tape winding	Prepreg tape
3.	Injection moulding	Reinforcement: staples Matrix: liquid
4.	Compression moulding	Reinforcement: staples/fabric Matrix: liquid/semi-molten
5.	Autoclaving	Reinforcement: fabric Matrix: liquid
6.	Thermoforming	Reinforcement: staples/unidirectional (UD) fabric Matrix: semi-molten
7.	Extrusion	Reinforcement: staples Matrix: liquid
8.	Pultrusion	Reinforcement: fabric/filaments Matrix: liquid
9.	Resin transfer moulding	Reinforcement: fabric/filaments Matrix: liquid

4.5.3 Fibre Dispersion and Orientation

Fibre dispersion is a phenomenon in short fibre-reinforced composites. *Fibre dispersion* is important for the complete wetting of the fibres in the composites and also for their orientation in the direction of the flow of the matrix. Both of these factors are important for creating the good properties of the product. Fibre dispersion is facilitated by the following factors:

- Absence of cohesive bonding among the fibres.
- No deterioration of surface treatment on the fibre surface. Surface treatment helps in easier handling and the processability of the fibres.
- Absence of moisture in the fibres. The moisture, being hydrophilic in nature, causes an incompatibility of the fibres with the hydrophobic resin.

The dispersion of fibres depends upon their density and on the flow rate of the resin. High-density fibres tend to settle down due to gravity. This phenomenon is seen in large products. The role of the temperature is also observed in the orientation of fibres because low temperatures favour high viscosity, and the fibre position in the product is frozen fast. The flow rate and hydrophobic surface of the fibre control part of the flow of the resin.

4.5.4 Troubleshooting

As it has been discussed in previous sections, rheology in polymeric composites has its significance in the fabrication/production of the product while obtaining the desired properties. The rheology in polymeric composites has many manifestations; therefore, its reasons and solution are discussed here.

1. *Viscosity of Resin*—High viscosities of resin do not allow it to flow and prevent the percolation of resin around the reinforcement. Therefore, poor wetting occurs, and this leads to resin-starved and resin-rich segments in the composites, as shown in ◘ Fig. 4.9a, b, respectively. Thus, the composite obtained does not have the requisite quality. Similarly, although low viscosities of the resin cause the reinforcement to be wet and produce good percolation of resin around the reinforcement, some time before the resin cures around the reinforcement it settles down due to gravity. Thus, the results of low viscosity cause resin-starved and resin-rich segments. Therefore, the viscosity of the resin must be optimum, and once resin has uniformly wetted the reinforcement, the gelation must take place so that resin remains at the right place to keep the volume fractions of the matrix and the reinforcement under control.

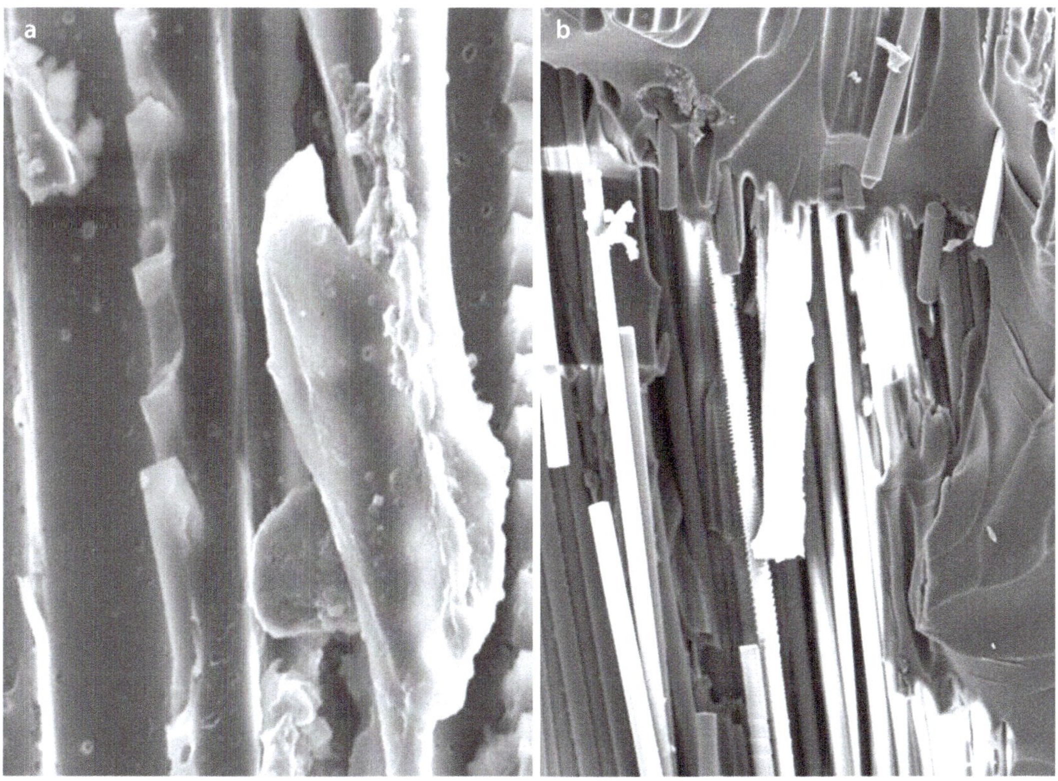

◘ **Fig. 4.9** **a** Resin-rich and **b** resin-starved segments in the composites [6]

2. *Temperature of Matrix*—The temperature of the matrix regulates the viscosity, gelation, and final curing. In fact, temperature controls the major portion of the curing/moulding cycle of the polymeric composites. Even the mould temperature plays a vital role in the fabrication of composites. If the temperature is not uniform throughout the mould or the composite part, the product obtained would have varied cross-linked densities. This would lead to a poor quality product. In order to obtain a uniform cross-linked density, it is important to soak the mould or product at the following temperatures:
   - Temperature at which minimum viscosity occurs
   - Temperature at which gelation occurs
   - Temperature in between the gelation temperature and the highest curing temperature
   - Highest curing temperature

   Demoulding at a high temperature may cause warpages. *Warpage* may be defined as the dimensional distortion in the composite product when it is demoulded. Warpages may be following five types: bow, crook, kink, cup, and twist. It is more common for thin sections of composites and for carbon fibre-reinforced composites. Carbon fibre is the only reinforcing commercial material that has a negative coefficient of thermal expansion. Therefore, carbon fibre goes under warpage quite easily, as shown in ◘ Fig. 4.10. The demoulding of all polymeric composites must be done at an ambient temperature in order to avoid warpages.
3. *Pressure on a Matrix*—Pressure and a vacuum have an influence over the flow rate of a matrix. Also, in resins where condensates are being evolved, the application of pressure and a vacuum removes all condensates. Otherwise, these conden-

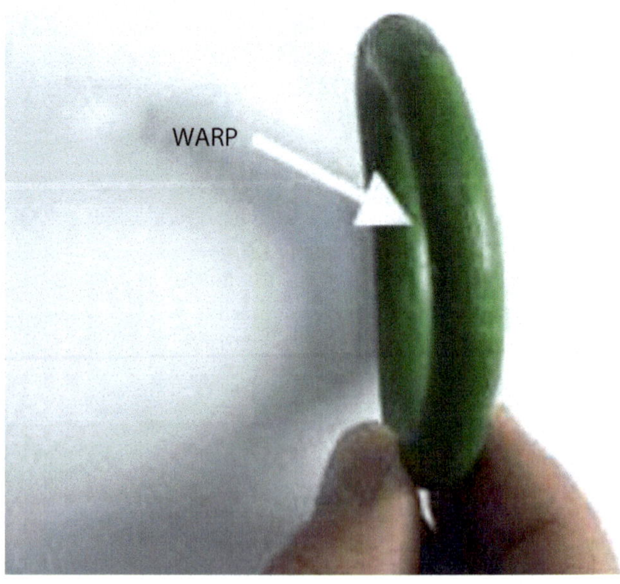

◘ **Fig. 4.10** Warpage in a thermoplastic composite [7]

sates would cause the improper curing of the composite, leading to delamination. In most of the processing technologies (such as compression moulding, injection moulding, matched die moulding, etc.), pressure ensures good compaction and dimensional accuracies in composite products. In injection moulding, the high injection pressure may cause overpacking of the mould, leading to warpages and dimensional inaccuracies in the moulding [8]. In resin transfer moulding (RTM), a vacuum and pressure cause wetting of the reinforcement, leading to proper impregnation.

4. *Raw Materials*—The primary raw materials for composites are reinforcement and matrix. The secondary raw materials are additives—if needed. Reinforcement must have a hydrophobic surface for its compatibility with resin. In case the reinforced is quite old and has not been kept under proper storage conditions, the coating of the coupling agent gets eroded. For example, if glass fibre is kept under high humid conditions, the coupling agent over it gets eroded. This could be experimentally observed by smearing a layer of resin over it using a paint brush, and it could be observed that the resin gets collected in patches without adhering to the surface of glass fibre evenly. Therefore, proper storage and handling will keep the reinforcement good for use. Similarly, the storage temperature and humidity for resin is important. The resin gets little cross-linked during a prolonged period of storage at ambient or high temperatures so that small lumps of resin may be seen in the storage drum. All of this occurs due to the advancement of a reaction, and the main cause for this advancement is an incorrect storage temperature followed by exposure to atmospheric air and moisture. The wrong temperature leads to an increase in viscosity over a period of time, and high viscosity and the formation of lumps do not allow the resin to flow and percolate as discussed.

   The role of an additive is also important in polymer processing. Additives either impart required properties or facilitate the polymer processing. Additives, such as particulate reinforcements (i.e. reinforcing carbon black, nanoparticles, etc.), improve the mechanical, thermal, and electrical properties of composites, whereas additives, such as diluents, lower the viscosity of highly viscous resins with some compromise to the overall properties of the composites [9].

4.6 Influence of Rheology on Composite Products

The effect of rheology on the quality and properties of polymeric composites has been generally described in the previous sections. Rheology is responsible for the flow of the matrix in the polymeric composite and also the movement of the additives and fillers that it carries. Therefore, rheology influences the final properties of poly-

4

Table 4.5 Rheological properties vs. final properties of polymeric composites

Serial No.	Rheological behaviour of matrix	Resultant effect	Final property of polymeric composites
1.	High viscosity	Restricted flow, poor wetting of fibres, resin-rich or starved areas	Poor overall properties
2.	Low viscosity	Smooth flow, good wetting of fibres	Good overall properties (provided care is taken that resin does not flush out of the mould)
3.	Quick solidification after wetting of fibre	Good wetting of fibres	High amorphous content in semi-crystalline polymers Reduction in T_g
4.	High elongational viscosity	Matrix retracts after release of pressure	High loss of modulus
5.	Sedimentation of filler or additives	Phase separation and segregation	Nonuniform properties
6.	Collection of matrix in patches on the fabric	Lack of compatibility between matrix and reinforcement	Delamination Poor overall properties
7.	Low viscosity due to solvent	Residual solvent and porosity	Low modulus Defective morphology and shrinkages Poor overall properties
8.	Viscosity build-up due to condensation and reaction of matrix	Porosity in the composite and resin-starved areas	Defective morphology Poor overall properties

meric composites. The effect of rheology on the final properties of polymeric composites is given in Table 4.5.

Apart from these effects of rheology, some important properties of composites that are influenced by rheology are explained next.

4.6.1 Product Shrinkage

As a matter of fact, all matrices shrink after curing or moulding; however, the degree of shrinkage may vary. Shrinkage also can be a processing aid because it helps in getting the mouldings attached on to the core of the mould. This makes it easy to remove because the peripheral ends of the moulding on the core of the mould provide a location to push the moulding out of the mould. In a reverse situation, expansion of the matrix will lead to the moulding becoming attached to the cavity of the mould. This would make it difficult to remove from the mould because the peripheral ends of the moulding would be inside the cavity [10]. However, minimum shrinkage is permissible for good product quality and properties. High shrinkages occur due to a high flow rate of the matrix during moulding.

4.6.2 Surface Properties

The surface properties are affected by the flow behaviour of the matrix. In short fibre composites, the alignment of the fibres occurs along the walls of the mould due to a shearing mechanism. The properties of the composites on the surface are superior in the longitudinal direction compared to the properties at the core of the moulding. The shear flow experienced at the core is the minimum flow; therefore, the orientation of short fibres is often random. The presence of solvent in the matrix causes pinholes and bubbles. The presence of moisture in the matrix leads to bubbles. The quality and properties of the surface of the moulding are directly dependent on the processability of the materials under desired conditions. High-temperature curing or moulding cycles are common with polymeric composites; therefore, in order to improve the processability of the polymeric composites, the raw materials must be dry so that surface or core defects do not occur. The presence of reactive fillers and additives must be avoided. Any impurity is a main hindrance in the processability. Impurities act as the nucleation point for surface and core defects in the matrix and polymeric composites.

4.6.3 Rheology in Structure–Property Correlation

High flow rates cause stretching and elongation of the polymeric molecules. These thermodynamic unstable polymeric chains have a high energy content and a high configuration of entropy. On curing, these molecules will relax and stabilize, attaining low entropy and causing shrinkage. This phenomenon also causes warpages and morphological irregularities in the polymeric composite.

Point to Ponder…
Viscoelasticity is a unique property of polymers that shows viscous behaviour in the form of flow and elastic behaviour in the form of rigidity simultaneously.

4.7 Rheological Measurement

Rheological measurement can be divided into three major categories: viscoelastic, steady shear, and elasticity. The difference between each of these tests is in the way the material is deformed.

4.7.1 Viscoelastic Measurements

Viscoelastic measurements reveal the flow and recovery behaviour of polymers and composites under stress and strain.

1. *Creep and Stress Relaxation*—Polymers are generally viscoelastic; hence, they exhibit creep and stress relaxation behaviour. Creep and stress relaxation tests measure the viscoelastic behaviour by monitoring changes in elongation at a constant stress and by measuring stress reduction at a constant elongation, respectively. Both tests permit the calculation of the same

4

viscoelastic constants. Most of the examples given here involves creep phenomena but can be applied to stress relaxation.

Several models have been developed to describe the viscoelastic nature of polymers. The models consist of Hookean springs and viscous Newtonian dashpots. A *Hookean spring* elongates instantaneously to an equilibrium elongation ε, which is directly proportional to the applied stress σ (i.e. $\varepsilon = \sigma/E$ where E is Young's modulus) and instantaneously recovers when the stress is removed. A *dashpot* is similar to a shock absorber in that an equilibrium elongation is never reached and it does not recover elongation already incurred. Mathematically, the applied stress σ is equal to the rate of elongation through a proportionally constant η known as the viscosity effect of an applied and removed stress as a function of time, i.e.

$$\varepsilon = \left(\sigma / \eta \right) t \quad (4.34)$$

There are three relatively simple models, viz. Maxwell, Kelvin–Voigt, and the four-parameter. The creep behaviour of the Maxwell model adequately represents the instantaneous elongation or recovery and permanent deformation due to the stress in a polymer. However, Maxwell model does not characterize the exponential decrease in the elongation rate or the recovery demonstrated by most polymers. The Kelvin–Voigt model is deficient in near instantaneous elongation or recovery. Also, this model reaches an equilibrium elongation that is not indicative of a true polymer and shows upon the permanent set.

The best features of both the Maxwell and Kelvin–Voigt models are combined into the four-parameter model. The creep response of this model adequately matches the creep behaviour of the distribution in molecular weight and other complex morphological structures associated with polymers. Thus, several models consisting of many Maxwell or Kelvin–Voigt elements with a distribution of relaxation times have been introduced, although the equation is complicated, as shown for a Kelvin–Voigt series:

$$J(t) = \int_0^a L(\lambda)\left[1 - \exp(-t/\lambda_1)\right] d\lambda \quad (4.35)$$

where J is the compliance and $L(\lambda)$ is the continuous relaxation time distribution.

2. *Dynamic (Oscillatory) Shear Rheology*—Dynamic measurements are very useful in terms of understanding the structure of delicate materials over a short and medium period of time. Dynamic measurements deal with the state of the material due to the quiescent structure at small deformations. These measurements provide valuable information regarding the microstructure of the sample(s) under investigation as well as their processability. Dynamic melt rheology usually involves imposing a small amplitude sinusoidal strain equation as

$$\gamma(t) = \gamma_0 \sin(wt) \tag{4.36}$$

and measuring the resultant sinusoidal equation as

$$\tau(t) = \tau_0 \sin(wt + \delta) \tag{4.37}$$

where
$\gamma(t)$ = Sinusoidal strain
γ_0 = Strain-amplitude
$w\omega$ = Frequency of oscillation
$\tau(t)$ = Sinusoidally varying stress
τ_0 = Stress amplitude
δ = Phase angle

Useful information that can be derived from dynamic shear rheology includes elastic (storage modulus, $G^{'}$), viscous (loss modulus, $G^{''}$), and complex viscosity ($\eta*$) equations, as given here:

$$G' = \left(\frac{\tau_0}{\gamma_0}\right)\cos\delta \tag{4.38}$$

$$G'' = (\tau_0 / \gamma_0)\sin\delta \tag{4.39}$$

$$\eta* = \sqrt{\left[(G'/w)^2 + (G''/w)^2\right]} \tag{4.40}$$

These viscoelastic parameters are directly related to the quiescent structure of the materials concerned.

The transition from liquid-like to solid-like natures of the unfilled and filled polymers can be analysed from the power law slopes of $G^{'}$ versus the frequency at low frequencies. This slope characterizes the quiescent nature of composites. A reduction in power law slopes (i.e. $\omega^2\ \omega^0$) for $G^{'}$ would tend to suggest the formation of a percolated network superstructure. $G^{'}$ is used as it is very sensitive to changes in the mesostructure of the material. The formation of such structures restricts the mobility of the polymer chains, thus enhancing the ability to store energy. This energy storage capacity is depicted as the solid-like response of $G^{'}$ at low frequencies.

4.7.2 Steady Shear Rheology

A steady simple shear flow is essential in applied steady shear rheology for two reasons:

- It is very easy to generate in the laboratory; hence, data most often reported are based on this flow.
- Many industrial processes (e.g. extrusion and flow in many types of die) approximate a steady simple shear flow.

1. *Shear Flows: Viscometric Functions*—Viscometric functions are defined for a steady simple shear and for a class of flows that are geometrically equivalent to steady simple shear. In a

steady simple shear, we have a velocity v_1 in the x_1 direction, which is a function of x_2 only where (x_1, x_2, x_3) are Cartesian coordinates of a point in the liquid. The shear stress is $\sigma = \sigma_{21}$, and the normal stresses are σ_{11}, σ_{22}, and σ_{33}. If the shear rate is $\dot{\gamma}\gamma$, which is defined to be $\partial v_1/\partial x_2$, then the viscosity ($\eta$) and the normal stress coefficient (Ψ_1, Ψ_2) are even functions of $\dot{\gamma}\gamma$ and are defined by

$$\eta(\dot{\gamma}) = \sigma(\dot{\gamma})/\dot{\gamma} \quad \textbf{(4.41)}$$

$$\psi_1(\dot{\gamma}) = N_1(\dot{\gamma})/\dot{\gamma}^2 = [\sigma_{11}(\dot{\gamma})]/\dot{\gamma}^2 \quad \textbf{(4.42)}$$

$$\psi_2(\dot{\gamma}) = N_2(\dot{\gamma})/\dot{\gamma}^2 = [\sigma_{22}(\dot{\gamma}) - \sigma_{33}(\dot{\gamma})]/\dot{\gamma}^2 \quad \textbf{(4.43)}$$

where N_1 and N_2 are known as the primary and secondary normal stress differences, respectively.

These functions may be obtained experimentally; $\eta(\dot{\gamma}\dot{\gamma})$ should be available for any polymer. $\Psi_1(\dot{\gamma}\dot{\gamma})$ may be available (at least for a similar polymer), but $\Psi_2(\dot{\gamma}\dot{\gamma})$ is not easily measured for polymer melts.

There are a number of expressions for the viscosity function that are used to fit empirical data and may be used in calculations. The most common of these is the *power law model*, which is given by

$$\sigma(\dot{\gamma}) = K\dot{\gamma}^n \quad \textbf{(4.44)}$$

or

$$\eta(\dot{\gamma}) = K\dot{\gamma}^{(n-1)} \quad \textbf{(4.45)}$$

where K and (to a lesser extent) n are temperature-dependent parameters.

For polymer melts, $0 < n < 1$ is generally observed (shear-thinning behaviour); n may be obtained from the slope of a graph of $\log \sigma$ against $\log \dot{\gamma}\gamma$. Generally, this will be adequate over restricted ranges of the variables and, in particular, will not be good for very small values of $\dot{\gamma}\gamma$ or σ where Newtonian behaviour ($n = 1$) is observed. An improvement that shows power law behaviour for large $\dot{\gamma}$ and Newtonian behaviour for small $\dot{\gamma}\dot{\gamma}$ is given by

$$\eta(\dot{\gamma}) = \eta_0 / \left[1 + \eta_0 / K\dot{\gamma}^{(n-1)}\right] \quad \textbf{(4.46)}$$

It is often asserted that there is also an "upper Newtonian" region (i.e. when at very high shear rates, the viscosity again becomes constant) at a value η_∞ which is less than η_0. Experimental evidence for this is difficult to obtain (because of flow instabilities), but the viscosity function proposed by Carreau is given by

$$\eta(\dot{\gamma}) = \eta_\infty + (\eta_0 - \eta_\infty)\left[1 + (\lambda\dot{\gamma}^2)\right]^{(n-1)/2} \quad \textbf{(4.47)}$$

This equation has additional features and is quite successful in fitting experimental data by choice into its four parameters.

4.7.3 Elastic Behaviour

The study of the elasticity of filled polymer melts has not received as much attention as steady shear viscosity, possibly due to experimental difficulties in measurements. For unfilled polymers, the first normal stress difference, N_1, extrudate swell, and a capillary flow entrance pressure drop are the common means of measuring elasticity. In steady shear experiments, N_1 is normally used as a measure of elasticity and is plotted as a function of shear stress. The elasticity of filled systems depends on various factors (such as filler size, shape, and concentrations). It has been found that near isotropic fillers like TiO_2, $CaCO_3$, carbon black, and glass beads show a decrease in elasticity when compared with the unfilled polymer and particulate size. However, isotropic fillers also showed that N_1 increases above that of the unfilled system when fibres are used as fillers. This is possibly due to hydrodynamic particle effect resulting from strong orientations in the direction of flow. For isometric fillers, the normal component of stress is zero, while the shear component has a finite value, hence lower elasticity. However, in the case of anisometric fillers (e.g. fibres), the normal stresses are higher, leading to higher elasticity.

To Be Remembered

1. The study of flow and/or deformation of any material (fluid/solid) is called rheology.
2. Reynolds numbers <2100 signify laminar flow. The Reynolds number for liquid resins and molten polymers is <100.
3. In the Weissenberg effect, a non-Newtonian fluid climbs up the stirrer rod when it is stirred.
4. Dynamic viscosity or the coefficient of absolute viscosity is a measure of the internal resistance offered to the fluid.
5. Planar Couette flow is the laminar flow in the space between two parallel plates where one is moving relative to the other.
6. Demoulding at high temperatures may cause warpages in composite products.
7. Creep is the permanent deformation at a constant stress.
8. Loss angle of loss tangent is the ratio of the loss modulus (G'') and the storage modulus (G').
9. Dispersion is a process where matter is divided into the smallest possible size and mixed in another substance. Whereas in distribution, matter is divided into a regular size and mixed in another substance.
10. Thermosetting matrices in solid form are called novolacs. Novolacs initially melt like thermoplastics with the application of heat and then solidify into a polymeric cross-linked network structure of thermosets.

Exercises

1. Define simple shear flow. Why is it necessary to use simple shear flow geometry for rheological measurements?
2. Use the power law equation and prove that polymer melts have shear-thinning behaviour.
3. Find the shear rate components of a flow through a tapered tube of Ellis model fluid.
4. Draw a τ vs. $\dot{\gamma}$ plot for the following:
   (a) A shear-thinning fluid having yield stress
   (b) A non-Newtonian fluid exhibiting zero shear viscosity
5. Sketch the velocity profile for the flow of a power law fluid between two horizontal plates where the upper plate is moving and the pressure gradient is given as
   (a) $\Delta P > 0$
   (b) $\Delta P = 0$, and
   (c) $\Delta P < 0$
6. Determine the shear rate components in (a) the axial annular flow with an imposed pressure gradient, (b) radial flow of the power law fluid between two parallel discs.
7. The rheological data of glass-reinforced polypropylene at 190 °C is reported as follows.

$\dot{\gamma}$ (s^{-1})	η (Pa.s)
0.1×10^{-1}	0.130×10^{5}
0.2×10^{-1}	0.129×10^{5}
0.1×10^{0}	0.125×10^{5}
0.1×10^{1}	0.900×10^{4}
0.1×10^{2}	0.500×10^{4}
0.1×10^{3}	0.110×10^{4}
0.1×10^{4}	0.100×10^{4}

Determine rheological parameters for this composite using the power law and Bird–Carreau models.

8. Discuss the following"
   (a) Normal stress measurement vs. the elastic nature of polymer
   (b) Elongational viscosity vs. stretchability
   (c) Relaxation time vs. product quality
9. Identify and justify the assumptions that are being used for the measurement of viscosity. List two geometries used for rheological measurements.
10. Derive the shear rate tensor for Couette flow between two concentric cylinders.
11. The flow behaviour of polymer melt shows at low and high shear rates what appear to be Newtonian viscosities with an increasing viscosity with shear rate in between. Name and compare at least two rheological models that can be used to predict this behaviour.

12. The viscoelastic behaviour of a certain polymer is to be represented by spring and dashpot elements having constants of 2 GN/m^2 and 90 GN/m^2, respectively. If a stress of 12 MN/m^2 is applied for 100 sec and then completely removed, compare the values of strain predicted by Maxwell and Kelvin–Voigt models after 150 sec.
13. A commercial sample of polyethylene has the following Ellis model parameters: $\eta_0 = 1.3 \times 10^4$ $N.s/m^2$, $\tau_{1/2} = 7.0 \times 10^3$ N/m^2, and $\alpha = 2.5$. Calculate the volume rate of flow in a pipe with an internal diameter of 3 cm, length of 20 cm, and a pressure drop of 5.0×10^6 N/m^2.
14. Determine the Ellis model parameters for the following set of data on polystyrene obtained at 180 °C.

$\dot{\gamma}$ (s^{-1})	τ (MPa)
0.5	35
1.5	105
15	414
150	820
500	1030

15. Determine the viscosity of the polymer at the wall of the tube ($d = 8$ mm and $L = 150$ mm) for a flow rate of 15 kg/hr. at a pressure drop of 17×10^5 Pa.
16. A polymer tubing plant operates at a temperature of 210 °C. Due to a melt instability problem, the processing temperature has been dropped to 195 °C. Estimate the expected percent of change in the viscosity of the polymer melt caused by this change (activation energy, 10 Kcal/mole).

References

1. Usman, Saeed, Kashif, Hussain, and Ghaus, Rizvi, "Characterization of glass fiber-reinforced high-density polyethylene," *Plastic Research Online*, SPE, 2 May 2014.
2. Cox, H.L., "The elasticity and strength of paper and other fibrous materials," *British Journal of Applied Physics*, vol. 3 (1952), p. 72.
3. Callister, W. D., Jr., *Materials Science and Engineering- An Introduction, Composites*, John Wiley & Sons, 6th edition, 2003, Chapter 16.
4. Matin, M.L., Daivis, P.J., and Todd B.D., "Comparison of planar shear flow and planar elongational flow for systems of small molecules," *Journal of Chemical Physics*, 113 (20) (2000), p. 9122.
5. May, Preston A. and Moore, Jeffrey S., "Polymer mechanochemistry: techniques to generate molecular force *via* elongational flows," *Chem. Soc. Rev.*, 42 (2013), p. 7497.
6. Bhandakkar, A., Kumar, N., Prasad, R.C., and Sastry S.M.L., "Interlaminar Fracture Toughness of Epoxy Glass Fiber Fly Ash Laminate Composite," *Materials Sciences and Applications*, 5 (4) 2014, p. 231.
7. *Warpage*, http://www.plastictroubleshooter.com/ThePlasticTroubleshooter/warpage.htm

8. Daniel, Isaac M. and Murphy, Walter P., Northwestern University, Departments of Civil and Environmental Engineering and Mechanical Engineering, Director, Center for Intelligent Processing of Composites.
9. MAX CLR-HP A/B catalogue on clear liquid resin, High Performance Epoxy Resin System, USA.
10. Wood warping, Wikipedia, https://en.wikipedia.org/wiki/Wood_warping.

Processability of Thermosetting Composites

A. K. Ghosh, M. Dwivedi, *Processability of Polymeric Composites*,
https://doi.org/10.1007/978-81-322-3933-8_5

The processability of polymeric composites depends upon various factors ranging from raw materials, coupling agents, additives present, processing method, mould design, processing parameters, surrounding environmental conditions, and post-fabrication handling. Uniformity in the properties of raw materials is essential for the desired processability of polymeric composites. It is quite common in reinforcements that their mechanical properties diminish once their useful life is expired. In glass fabrics, it is also seen that wet resin does not stick to its surface, indicating the loss of the coupling agent on glass fibres, thus making them hydrophilic [1, 2]. The presence of additives influences the properties of thermosetting composites, including curing of the matrix, interphase with the reinforcement, etc. The right selection of the processing method and its parameters aids in processability. Mould design and tooling too have an influence on the efficiency and quality of thermosetting composites being manufactured. The surrounding environmental conditions (such as humidity, dust, and temperature) have an influence on raw materials and processing conditions. Handling a manufactured part requires care so that further processing and finishing operations (such as printing, embossing, metalizing, etc.) can be carried out.

These factors influence the processability of thermosetting composites, and the following sections delve into more details for a better insight on the processability of thermosetting composites.

5.1 Interface and Interphase in Thermosetting Composites

Interface is the surface of the adherent (i.e. reinforcement or matrix) that comes in contact with each other. The *interphase* is the continuous region of dissimilar composition between the reinforcement and the matrix, thus bonding them. Interfacing is bonding the surface between the reinforcement and the matrix and across where a discontinuity in some parameter occurs. Due to dissimilarity in the surface properties of the reinforcement and the matrix, the region between them forms the third phase in the composites, which is called the interphase. An interphase is the region through which material parameters (such as crystal structure, atomic structure, tensile modulus, density, coefficient of thermal expansion, etc.) change from one side to the other. The chemical composition of the interphase is such that one end of interphase binds with the reinforcement and the other end binds with the matrix. One end of the coupling agent makes a covalent bond with matrix, while the other end forms a hydrogen bond with the inorganic reinforcement. The hydrogen bond remains labile and does not break on experiencing minor strains during a fibre-to-fibre load transfer through shear in composites. In composites, a fibre-to-fibre load is transferred through the matrix. Therefore, the matrix experiences shear force during the load transfer. The classical theory of interphase advocates that the bond between the reinforcement and the matrix is strong and rigid [3–5]. In such a situation, the interphase bond fails on the slightest strain and, thus, causes failure of composites during

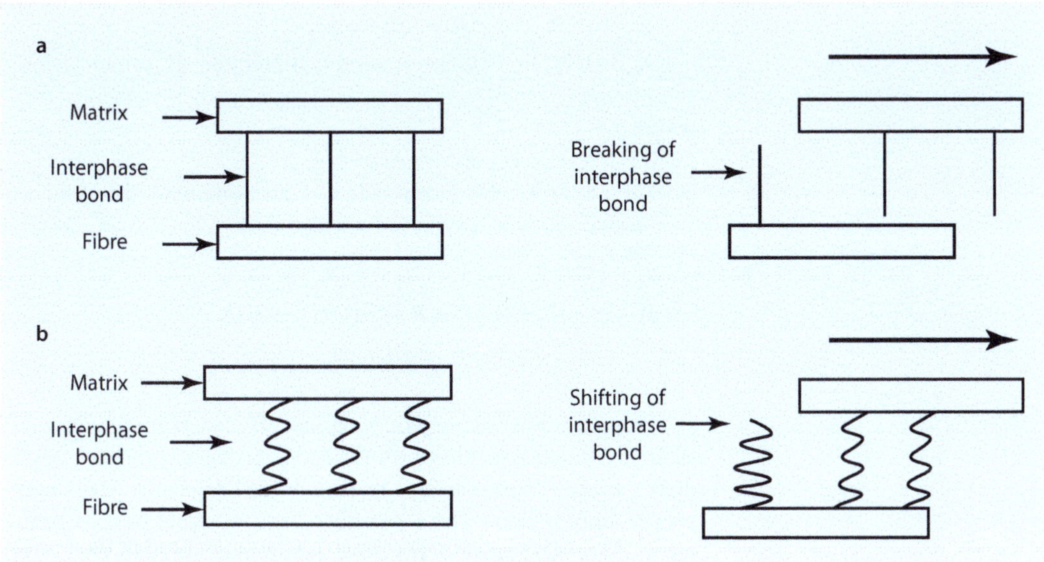

Fig. 5.1 **a** Classical theory of interphase formation and **b** Pludeman's theory of labile linkages of interphase [6]

load transfer. This has been shown schematically in Fig.5.1a. A new theory of interphase has been proposed by a noted scientist named E. Pludeman. The schematic representation of Pludeman's theory is given in Fig. 5.1b. According to this theory, there is a labile link between reinforcement and matrix through interphase. The interphase region forms a secondary bond, such as a hydrogen bond, with the reinforcement and a covalent bond with the matrix. This results into a labile linkage during high strains at the application of stress. This labile link causes the shifting of the interphase bond position, as shown in Fig. 5.1b. That is how the interphase does not fail during a load transfer in composites. This theory is well accepted among the composites' fraternity [6].

The most important fibres used in composite industries are glass fibres, carbon fibres, and aramid fibres (popularly known as Kevlar fibres). The mechanisms for their interphases with the matrix in composites are summarized here.

5.1.1 Glass Fibres

Glass fibres are not handled in their nascent condition. Due to their high hardness and brittleness, they break—even with the minutest of strain. Therefore, sizing is applied over the glass fibre surface. Sizing (e.g. polyvinyl alcohol (PVA), starch gum, hydrogenated vegetable oil, etc.) is applied on glass fibres to make them easier to handle so that glass yarn and fabric can be made. Subsequently, sizing is washed away from the fabric or rovings. Then an aqueous solution of silane coupling agent is applied on the surface of the glass fabric or rovings. Coupling agents (such as organosilanes or organotitanates) are used for adhesion of the matrix to the fibre, as described next.

$$Ri - Si - X_3 \quad [\text{Organosilane}]$$
$$X_3 Si(CH_2)Y \quad [\text{General chemical formula of organosilane}]$$

where

- R is a resin compatible group
- X is a hydrolyzable group which in turn interacts with glass surface aqueous solution of organosilane

5

$$R - Si - X_3 + 3H_2O = R - Si(OH)_3 + 3HX$$
$$(\text{Trihydroxysilanol})$$

The common silane coupling agents are trimethoxy(vinyl) silane, trimethoxy(chloropropyl)silane, trimethoxypropyl glysidoxy silane, trimethoxypropyl(methacrylate)silane, triethoxypropyl-amino-silane, etc. Glass fibre has entrapped oxygen in the form of an -OH group. This -OH group makes a hydrogen bond with an -OH group of hydrolysed silane. This hydrogen bond is in dynamic equilibrium and shifts when experiencing shear force, as shown in Fig. 5.2. The labile nature of the hydrogen bond is

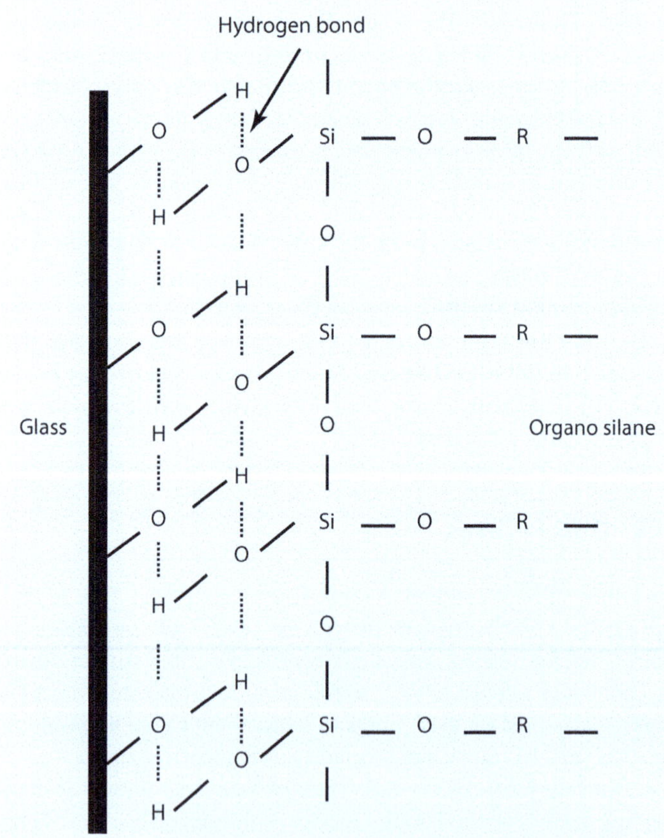

Fig. 5.2 Hydrogen bonding between glass fibre and organosilane

explained in Pludeman's theory. Note that –R groups make covalent bonds with matrix resin.

Titanate coupling agents are limited to inorganic reinforcements, and they form an organic monolayer over the surface of inorganic reinforcement, such as glass fibres. The usage of titanate coupling agents causes a reduction in the viscosity of the matrix after interaction with it. Titanate coupling agents cause good adhesion between fibres and the matrix, improve the dispersion of short fibres or fillers in the matrix due to reduction of the viscosity of the resin matrix, improve the impact strength of the composite, and do not cause embrittlement of the composite. Titanate coupling agents may be classified as monoalkoxy (isopropyl triisostearoyltitanate, isopropyl triacriloyltitanate, etc.), chelate (dimethacryloyl, oxoethylene titanate, di(dioctyl)pyrophosphate, oxoethylenetitanate), and neoalkoxy (neopentyl(diallyl)oxytrineodecanonyltitanate) coordinate and quat types [7].

Example 5.1 Explain the process of the application of a silane coupling agent on the surface of glass fabric in the industry.

Answer

The steps involved in treating the surface of glass fibres with silane coupling agent are given here.

- Step 1. A sizing agent such as starch, gum, or polyvinyl alcohol (PVA) is applied on the surface of freshly drawn glass fibres. This helps in the handling of glass fibres without breakages.
- Step 2. The sized glass fibres are converted into tarns, and they are woven into the fabric of the desired weave.
- Step 3. The sizing agent is washed with water to expose the nascent surface of the glass fibres.
- Step 4. The glass fabric is immediately passed through a bath containing an aqueous solution of a silane coupling agent. The speed of drawing glass fabric is kept constant in order to allow the silane to adhere reactively on the surface of the glass fibres.
- Step 5.The fabric is dried using a hot air blower and packaged for transportation.

5.1.2 Carbon Fibre

Carbon fibre contains pure carbon, and its fabric is shown in Fig. 5.3. The salient features of carbon fibres are the following:

- Carbon fibre is highly inert.
- The surface layer of nascent carbon fibre is weak due to skin-core heterogeneity.
- There is a difference in morphology between high-modulus (HM) and high-strength (HS) carbon fibres.

Fig. 5.3 Picture of woven carbon fabric [9]

- Carbon fibres are of various types (e.g. high strength, high modulus, ablative, etc.) [8].
- Carbon fibres are made from different precursors, such as polyacrylonitrile (PAN) fibres, mesophase pitch, and rayon fibres.

Due to the inert behaviour of carbon fibres, the surface of carbon fibres is activated by the following methods:
- Liquid-phase oxidation (bromine water)
- Gas-phase oxidation (mixture of oxygen and nitrogen)
- Chemical vapour deposition (CVD) in the case of carbon whiskers

The activated surface is coated with a very thin layer of either polyimide, epoxy, polyurethane, or phenolic resins. The polyimide or epoxy layer covalently bonds with activated carbon surface as well as with the matrix. The strain does not cause failure of the interphase due to the presence of the polymeric interphase, which has the capability to yield.

5.1.3 Aramid Fibres

Aramid fibres have a peculiar morphology and distinct low surface energies. The salient features of the morphology of aramid fibres are the following:

- Aramid fibre has highly oriented chain structure.
- It has a skin–core heterogeneity; hence it has a low interfacial strength.
- Poor adhesion with most polymers.

The methods to improve fibre/matrix adhesion are the following:

- Mechanical roughening of the surface.
- Silane coupling agent used for glass fibre.
- Isocyanate-linked polymers.
- Acid or base hydrolysis yields a terminal amino group on the fibre surface.
- Bromine water treatment.
- Plasma treatment in vacuum, ammonia, or argon.

The properties of the composite materials are determined by those of the raw materials, shape of the filler, morphology of the system, and nature of the interphase between the phases. The properties of composites are altered by a change in morphology or interphase. An important property of the interphase that can greatly affect mechanical behaviour is the strength of the adhesive bond between the fibre and the matrix [10, 11].

5.2 Physical Forms of Reinforcements

Mostly, reinforcements are in the form of fabrics. Fabrics are used in high-end applications where high mechanical properties are required. Reinforcement in the form of rovings and tows is used for high-performance applications, such as pressure vessels using filament winding process. The common forms of fabric reinforcements are given next.

Woven Fabrics Woven fabrics provide continuous reinforcement to composites, and their weaves and patterns have been discussed in ▶ Chap. 1. Fabrics made out of glass fibres, carbon fibres, graphite fibres, Kevlar fibres, PEEK fibres, UHMWPE fibres, etc. are available.

Chopped Strand Mat (CSM) CSM has been discussed in detail in ▶ Chap. 1. It is commonly made out of glass fibre staples.

Meshes Meshes are made out of both glass and carbon fibres. Their thickness is usually small (approximately 7–10 μm and 15–20 μm for carbon fibre meshes and glass fibre meshes, respectively). They are used for the surface finishing operation during the moulding of composites.

Staples/Short Fibres Staples/short fibres are available from most of the materials, such as glass fibres, carbon fibres, Kevlar fibres, etc.

Particles Particles are regular-shaped, small-sized entities. They form a large range of materials. They are of two types: reinforcing and non-reinforcing. The most common particles are carbon black, silicon dioxide, talc, calcium carbonate, and titanium dioxide. Mostly particles are used for mixing in the composites as the fillers.

Whiskers Whiskers are irregular-shaped, small-sized entities. They have high mechanical properties and specifically have a high modulus. Whiskers have reinforcing properties, but their materials are limited. The most common whiskers used are silicon carbides, aluminium oxide (alumina), and aluminium carbides.

Natural Fibres Natural fibres are gaining importance due to environmental concerns. 'Green composites' are fabricated using natural fibres (such as hemp, sisal, jute, coir, flax, wood fibre, etc.) as the reinforcement either with conventional polymeric matrices (such as epoxy resin, polyester resin, phenolic resin, etc.) or with biodegradable polymers (such as polylactic acid (PLA), polycaprolactone (PCL), polyglycolic acid (PGA), etc.). Natural fibres have the following advantages:
- Low-to-moderate density
- Good acoustic absorbance
- Low cost
- No irritation to the human body

Natural fibres have the following disadvantages:
- Hydrophilic in nature.
- Natural fibres of plant origin are susceptible to fire.
- Low-impact strength.
- Environmentally dependent qualities

Natural fibres have found many applications, such as door panels, back and front panel of car seats, etc., in the automobile sector.

5.3 Thermosetting Composites

Thermosetting matrices are the workhorses in the composite industries. Most of the high-performance polymeric composite structures are made using thermosetting matrices. The applications of thermosetting composites range from fuselages, airframes, nose cones, re-entry surfaces, nozzles for aerospace structures, automobiles, railway coaches, wind turbine blades, chemical storage tanks, armoured panels, and civil structures to air conditioning ducts, furniture, and sports goods. They are fast replacing conventional structural materials [12]. The most important features of thermosetting matrices enabling their wide uses are the following:
- Ease of processability
- Good fibre wetting due to low viscosities

Table 5.1 Attributes of thermosetting matrices

Serial No.	Parameters	Thermosets
1.	Shelf life	Short
2.	Fabrication time	High
3.	Post formability	Impossible
4.	Generation of byproducts	Yes
5.	Tooling	Inexpensive
6.	Mixing of chemicals	Required
7.	Production rate	Slow
8.	Economy	Low
9.	Handling	Not easy
10.	Recycling	Not possible

- Tailorable curing time
- Adaptability to most of the moulding processes
- Processability through different moulding methods and mould materials

The attributes of thermosetting matrices are provided in Table 5.1.

5.3.1 Curing of Thermosetting Composites

The most critical parameters for thermosetting composites involve an optimum curing cycle so that the composites are neither undercured nor overcured. The optimum curing provides the right combination of T_g and mechanical properties, allowing a composite to perform the best in the operational environment for which it is designed. The cure cycle determines the cross-link density too.

The *cure cycle* is a time and temperature combination to cross-link the thermosetting resin with optimum properties. Determination of the cure cycle is a laborious and time-consuming process. Most of the time, the cure cycle is prescribed by the manufacturer of the resin, and it must be noted that the cure cycle may differ slightly from batch to batch. In cases of high-performance and critical products, the cure cycle must be determined very carefully [13, 14].

In order to determine the cure cycle of the thermosetting resin, it is necessary to carry out a *differential scanning calorimetry* (DSC) of the resin. In a DSC curve, it can be observed that when the resin starts to gel with the onset of curing, energy is released due to an exothermic reaction. A typical DSC curve for the curing of thermosetting resin is shown in Fig. 5.4.

The peak of the DSC curve determines the maximum curing temperature for the resin system, and the area under the peak that

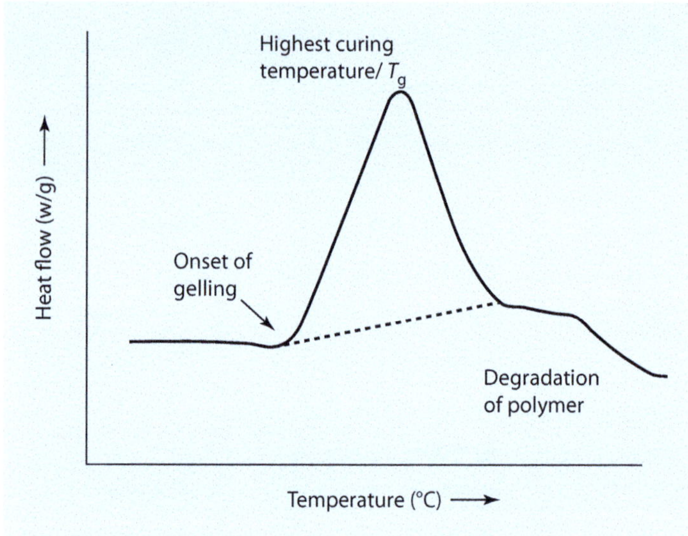

Fig. 5.4 Typical DSC curve of curing of thermosetting resin

is marked with dotted lines gives the measure of the exotherm in the curing of the resin. In other words, this gives the total heat content, H, of the resin during curing.

For an optimum cure, it is essential that the area under the peak be the maximum, and this exercise leads to a determination of the correct cure cycle for that resin. For this, the DSC is run at a constant temperature for the same resin against time. Four different temperatures with ±6 °C of the peak temperature are taken. If the peak temperature is T°C, the four temperatures are decided as $T + 3$, $T + 6$, $T - 3$, and $T - 6$ °C. The range of temperature may be increased based on the requirement of accuracies for the curing cycle. The DSC is run at these temperatures separately at constant heating rate, preferably 10 °C per minute. During these experiments, the peak temperature, at which the area under the curve is maximum, is taken as the curing temperature. The time of curing may vary depending upon the size or volume of the product and the type of mould used. It is calculated based on the heat and mass flow that 40 W of power is required for heating a 1 Kg of mass of steel mould. This procedure is followed as a thumb rule in most composite industries.

In the development of the cure cycle, the thermal equilibrium must be maintained for every important stage of the curing. Therefore, enough soaking time must be provided for the mould and component to reach equilibrium. The important stages are reaching the gelation temperature, the path from the gelation temperature to the peak curing temperature, and realizing the curing temperature. On the path between gelation and curing, the temperature must be held constant at two temperatures (reasonably spread out) in order to achieve thermal equilibrium. Otherwise, the cross-link density would vary due to the formation of 'hot' and 'cold' regions in the mould and component.

Fig. 5.5 Flexible add-on curing device [16]

Cross-link density is the direct measure of the extent of the curing of thermosetting resin. The higher the cross-link density, the better is the curing. According to the definition, the cross-link density is the concentration of chemical bonds within a cured polymer. When polymer chains are linked together by cross-links, they lose some of their ability to move as individual polymer chains. High cross-link densities cause polymers to become very rigid or glassy. It is measured in terms of hardness and the modulus of the cured component. A method to measure the cross-link density is to place the cured part in a solvent, and the weight of the part that is not dissolved or swollen in the solvent gives the indirect measure of the cross-link density. With an optimum curing temperature and curing time, the cross-link density is not only good, but it also is uniform [15]. The uniformity of the cross-link density is important for uniform performance of the composite product.

New curing methods are based on flexible add-on curing devices that provide out-of-oven curing solutions, as shown in Fig. 5.5. This technology is helpful for curing large composite products. This device is based on *carbon nanotube* (CNT) film that can heat and solidify a composite without using massive ovens. When this device is wrapped over multilayer composites and connected to an electric power source, the heated film stimulates the matrix to cure.

Curing is a complex chemical reaction in actual process. Post curing is essential to achieve complete curing along with high and uniform cross-link density. Polymer chains are cross-linked by 3-D bonding structure in the matrix. For example, in the case of polyester resins, curing takes place through styrene. In epoxy it takes place

5

through amines or anhydrides. Only when all linkable sites in the matrix are connected the reaction is said to be complete. When linkable sites are still present or unreacted styrene is still present in polyester, it is said that reaction is not complete. Therefore, further heating is required to complete the curing. This subsequent heating to complete the cure is called *post curing*.

Example 5.2 Explain the significance of the peak in a DSC curve

Answer

The peak in a DSC curve signifies the following:

- It indicates at what temperature maximum heat flow occurred in the polymer. Maximum heat flow is the value at which maximum exotherm occurred due to the optimum reactivity causing cross-linking.
- The peak indicates the average temperature between the curing regimes of the cure cycle.
- The temperature associated with peak is the cure temperature. Below this temperature, if curing takes place, the degree of cross-linking will not be optimum.

5.3.2 Selection Criteria for Thermosetting Matrices

The selection criteria for thermosetting matrices are dependent on the processability factors of resin as well as processing technology. A summary of processability factors already has been outlined in ▶ Chap. 1 (▶ Sect. 1.3). The processability factors for thermosetting resins are detailed here.

1. *Properties Driven by Selection Criteria*
   - *Viscosity*—The viscosity of resin during processing has an important effect on maintaining the even volume fractions of resin and reinforcement. Low viscosities help in processes involving low pressures like vacuum-assisted resin transfer moulding. This in turn helps in the reduction of shear forces experienced by resin and reduces the possibility of disturbing the reinforcement layup orientation. Therefore, low viscosities or low viscosities at high temperature (below the gelation temperature) are preferred for processing. Resins having low viscosities are preferred for the resin selection criteria.
   - *Wettability of Reinforcement*—The processability of resin only becomes improved with its capability to wet the reinforcement. Resin must have the capability to percolate around every fibre and particle. The surface tension of the resin must be low so that it smears the surface of the fibre, as shown in ◘ Fig. 5.6. Low wettability of the resin results in resin-rich and resin-starved areas in composites.

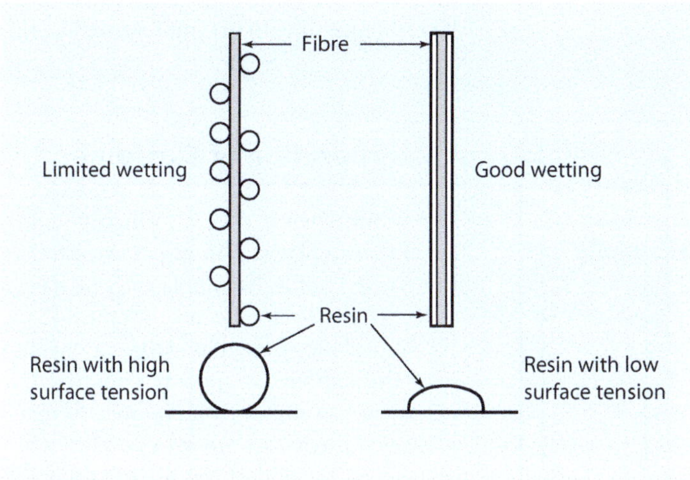

Fig. 5.6 **a** Limited wetting of fibre by resin having high surface tension and **b** good wetting of fibre having low surface tension

- *Chemical Interaction with Reinforcement*—Chemical interaction between the resin and reinforcement for making a chemical bond is preferred. The interphase of the composite depends upon the reactivity of resin as well as the reinforcement for each other. However, coupling agents on the reinforcement play a major role in this. Normally, thermosetting resins are reactive to the coupling agent on reinforcements. The resin must be selected based on its reactivity with the surface of the reinforcement to make a good interphase.
- *Curing Reaction*—Resins having additional reactions (such as epoxy resins, polyester resins, etc.) are preferred for composite processing. Resins undergoing condensation reactions, such as phenolic resins, give byproducts in the form of water, and PMR–PI resins give byproducts such as methanol/ethanol and N-methylpyrrolidone/dimethylacetamide. This creates multifold challenges during moulding of composites, such as the extraction of byproducts, consolidation of matrix, correct volume fractions, etc.
- *Molecular Weight and Its Distribution*—The high molecular weight of resins restricts their ability to flow and thus their percolation around the reinforcement. In the case of epoxy resins, there are bifunctional resins, such as diglycidylether of bisphenol 'A' (DGEBA), which has molecular weight of 30,000–70,000 g/mol and a viscosity of 10,000–12,000 cps. This viscosity comes down to 5000–7000 cps after mixing with a hardener (i.e. amines or anhydrides). The molecular weights of tri-functional and tetra-functional epoxy resins are in the range of 50,000–120,000 g/mol and have viscosities in the range of 20,000–50,000 cps. These are hot melt processed resins. The molecular weight distribution is quite narrow at approximately 1.5 for epoxy resins.

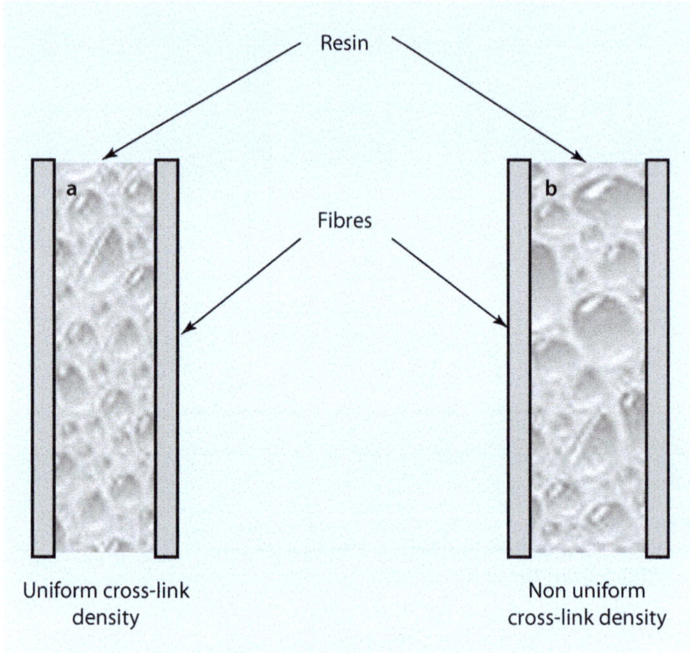

Fig. 5.7 **a** Uniform and **b** nonuniform cross-link density of resin

- *Cross-Link Density*—The cross-link density of a thermosetting resin governs the property of the composite. Uniform cross-linking throughout the matrix gives uniform properties. Uneven cross-linking occurs primarily due to the uneven curing of the resin or the presence of feebly cross-linked moieties in the resin that occurred due to the expiration of the resin life or improper storage conditions. Uneven cross-linked density has different sizes of molecular networks in the resin, which grossly affects the properties of the composite, as shown in Fig. 5.7. The resin matrix must be checked for its consistency before processing the composite so that the cross-link density is uniform.
- *Gelation Time*—This is one of the most important parameters from the point of view of processing. This parameter provides the processing window for making composites using any method and technology. The gelation time is predominantly the resin's property, and it is affected by the catalyst used; temperature of processing; reaction—if any—of the reinforcement; etc. Depending on the processing technology used, resin with an optimum gel time to provide the maximum processing window must be selected. For example, a hand layup process is manual and slow; thus, it requires large processing window. Therefore, resins with a high gelation

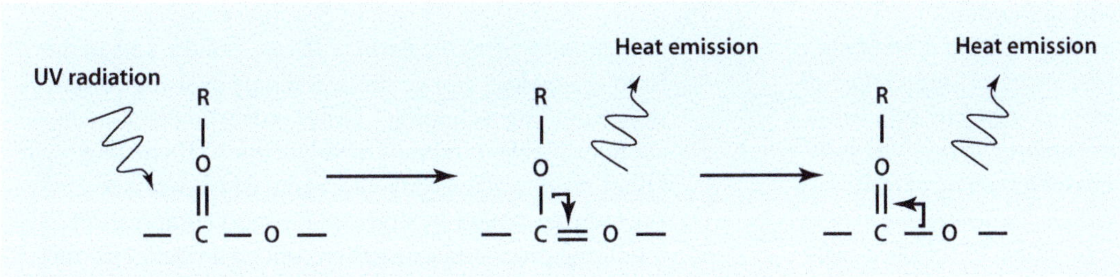

Fig. 5.8 Dissipation of UV energy as heat energy through the oscillation of a double bond between - C = O and - O —

time (e.g. polyester and epoxy resins) are suitable for using a hand layup process.

- *Thermal Stability*—The thermal stability of a resin is another criterion for its selection. The thermal stability of a resin governs the life of the composite under the required service conditions. The curing temperature of thermosetting resins is usually close to their glass transition temperature (T_g). Thermal degradation of thermosetting resin begins beyond its T_g.
- *Hygrothermal Stability*—The hygrothermal stability of the resin ensures its life under heat and in humid conditions. The ester group of polyester gets hydrolysed under humid and hot conditions. For example, polyester resins are not suitable for a hygrothermal environment, whereas epoxy and vinyl ester resins are more suitable.
- *Post-processing Operations*—The role of the matrix and its chemical activity is important for post-fabrication operations, such as metalizing, colouring, assembling, packaging, etc. Resin plays a role in moulding surfaces; therefore, processing technologies need to be chosen in order to ensure minimum post-fabrication operations on moulded composite products.
- *UV Resistance*—The selection of resin is heavily based on the operational environment. UV radiations have been a major cause of concern for resin systems because they readily degrade under UV exposure. In polyesters, UV energy is absorbed at double bond of –C = O and dissipate UV energy as heat energy through the oscillation of a double bond between –C = O and –O, as shown in Fig. 5.8.
- *Fire Resistance*—Resistance to fire is one of the most important criteria to selecting resin matrices. Epoxy and polyester resin-based composites are susceptible to fire hazards. Composite products that are exposed to a high heat environment, such as fire retardants, have bismuth oxides or antimony oxides added to improve their fire resistance. Phenolic resin-based composites are inherently fire resistant.

Point to Ponder...
Glass CSM-reinforced polyester shades are used for outdoor applications due to UV resistance of resin.

Point to Ponder…

Novolac *is a solid form of resin which melts at application of heat and, therefore, behaves as thermoplastic initially. Its melting helps in wetting of reinforcement. On further heating at high temperature, it solidified with cross-linked network of thermosetting resin. Novolacs are usually used as prepregging resins.*

2. *Processing Technologies Driven by Selection Criteria*
   - *Compatibility with the Form of Resin*—Solid resins in the form of novolacs can be processed only by using compression moulding technology. Other processing technologies are not suitable to process novolacs due to their solid form. Other resins, such as phenolic resin, which releases byproducts, cannot be processed using a compression moulding process because there will be no outlet for the byproducts. Rather, the byproducts will get entrapped in the moulding, causing a defective composite.
   - *Flow of Matrix*—The processing technology must ensure the flow of matrix into the entire moulding cavity. Flow of matrix is also dependent on the design of the mould and the reinforcement used. All of the moulding processes support a liquid form of the resin during processing.
   - *Compatibility of Mould Design with the Matrix*—Process compatibility depends on both the mould design and material used for the mould. For high-temperature curing resins, only metallic moulds are used. For room-temperature curing, resins (such as polyester and epoxy resins), plaster of Paris moulds, wooden moulds, etc. are used in addition to metallic (steel and aluminium) moulds due to their low cost.
   - *Production Rate*—Resins that have a short processing cycle are used to achieve high production rates. The curing of polyester and epoxy resins may be quickened by slightly increasing the content of the catalyst and the hardener, respectively. However, this content cannot be increased indiscriminately because excessive exotherm may damage the composite. A continuous section of composites having a regular cross section can be processed through pultrusion. Similarly, a cylindrical product, such as a high-pressure vessel, may be processed using a filament winding method. These examples prove that the processability of composites is process-dependent too. The selection of process ensures the viability of the composite product. There might be two or more processing methods that can be used for a composite product, but an engineering analysis in terms of performance, tooling cost, processing cost, raw materials, etc. would decide upon a single process only. In other words, the product can be processed using one specific method only.
   - *Cost of Production*—Resins are selected to reduce the cost of production. Depending upon the service environment and expected service life of the composite product, a resin is chosen to reduce the cost of production of composites. For example, polyester resin has three grades which are (i) *general-purpose polyester resin* that is very low in cost coupled with low mechanical properties, high shrinkage, and low transparency; (ii) *isophthalic polyester resin* that has a moderate cost coupled with good mechanical properties, low shrinkage, and high transparency; and (iii) *vinyl ester*

Fig. 5.9 High-pressure hydrogen fuel vessel (type IV) made out of a carbon fibre-reinforced filament-wound composite [17]

resin that has a high cost coupled with excellent mechanical properties, very low shrinkage, and high transparency. Depending upon the product and service requirement, resin may be chosen.

- *Product Size and Shape*—The shape and size of the product limits the processability of composites. An irregularly shaped product having undercuts on its profile is difficult to mould using a filament winding process. Similarly, large composite sheets are not processable by injection moulding, resin transfer moulding, tape winding, or filament winding. The wrong selection of the processing method leads to a poor-quality product with a high overall cost of production. In order to process any composite product, the processability of the product depends upon the procedure selected with respect to the product's shape and size. For example, to manufacture a pressure vessel, the most suited processing method is a filament winding process. The Maruhachi Co, Ltd. of Japan has developed a high-pressure hydrogen fuel storage vessel (type IV) with a carbon fibre-reinforced filament wound structure and an impermeable plastic lining, as shown in Fig. 5.9.

5.3.3 Processability of Commonly Used Thermosetting Matrices

Thermosetting resin solidifies or cures through exothermic reactions. Some of the commonly used thermosetting resins and their processability factors are described next.

1. *Epoxy Resins*—Epoxy resins are the workhorse of the composite industry. They are used in most high-performance applications of composites. They are manufactured through condensation reactions. They are available in various grades, such as bi-, tri-, and tetra-functional resins. The most common epoxy resin, which is popularly known as Araldite® and also is

used as adhesive, is diglycidyl ether of bisphenol 'A'. It is manufactured from the condensation reaction of bisphenol 'A' (BPA) and epichlorohydrin (ECH) in the presence of sodium hydroxide. The condensate comes out in the form of sodium chloride and water. The resin is used in the form of a prepolymer. Fluoro-epoxy resins are used for fire-resistant applications. Elastomer-toughened epoxy resins are also available, and they improve the toughness of the composite product. The general properties of epoxy resins have been mentioned in ▶ Chap. 1. The major processability factors for epoxy resins are as follows:

- The viscosities of epoxy resins can be controlled by diluents.
- Room-temperature curing hardeners, such as diethylene triamine or triethylenetetramine, provide a processing window of approximately 10 minutes, which is its gelation time.
- High-temperature curing hardeners, such as diamino diphenyl sulfones (DDS) or diamino diphenyl methane (DDM), are cured at 160–180 °C and provide a long processing window of 2–3 hours.
- Epoxy resins have low shrinkage (<0.5%).

2. *Unsaturated Polyester (UPE) Resins*—There are two types of unsaturated polyester resins. One is a general-purpose resin made from a condensation reaction of phthalic anhydride, propylene glycol, and maleic anhydride. The other is an isophthalic resin made out of a condensation reaction of acid, propylene glycol, and maleic anhydride.

   An isophthalic grade of UPE has superior properties over a general-purpose UPE. These resins are mixed with α-methyl styrene (which is cheaper in cost than styrene) to make the resin in a liquid form. For curing, an initiator, such as methyl ethyl ketone peroxide (MEKP), and a catalyst, such as cobalt naphthenate, are mixed in the resin. Benzoyl peroxide along with a tertiary amine is also used as an initiator for the curing of UPE resins at room temperature. The catalyst is mixed after the initiator is mixed in the resin; otherwise an excessive exothermic reaction may lead to an explosion. The cross-linking and curing reactions in UPE resins are governed by free radical polymerization. During a cross-linking reaction, the double bond of maleic anhydride gets saturated with styrene. This makes the cross-linked network with the styrene. The general properties of UPE resins have been mentioned in ▶ Chap. 1. The major processability factors for UPE resins are given here.

   - The molecular weight of UPE resin is in the range of 1200–3000 g/mol.
   - The viscosities of UPE resins are low.
   - UPE resins are curable at room temperature only. The processing window is approximately 10–15 minutes, which is its gelation time.
   - UPE resins have high shrinkage (approximately 1.5%).

3. *Vinyl Ester Resins*—Vinyl ester resins are made through a condensation reaction of DGEBA with acrylic or methacrylic acid. Methacrylic acid-based vinyl ester resins are used in the composite industry, whereas acrylic acid-based vinyl ester resins are used for coating purposes. Epoxy resin is commonly used for manufacturing of composite structures; therefore, the properties of vinyl ester resins may be tailored for the use of different grades of epoxy resins. Water from a reaction comes out as a condensate. Vinyl ester resins have properties of both epoxy and polyester resins. Vinyl ester resins have both good mechanical strength (like epoxy resin) and good UV resistance (like UPE resin). The curing agents, curing reactions, and mechanisms are similar to those of UPE resins. The general properties of vinyl ester resins have been mentioned in ▶ Chap. 1. The major processability factors for vinyl ester resins are given here.
   - The viscosities of vinyl ester resins are low.
   - Vinyl ester resins have low shrinkage (> 0.5%).
4. *Phenol Formaldehyde Resin*—Phenol formaldehyde resin is commonly known as phenolic resin. It is made from a condensation reaction of phenol and formaldehyde. This goes through a further condensation reaction during curing and releases water from the reaction as a condensate. Phenol formaldehyde resin is used as a matrix for high-temperature applications, ablative liners (due to good charring capabilities), and the fabrication of carbon–carbon composites (CCC) (again due to good carbonaceous content and charring capabilities). The general properties of phenolic resins have been mentioned in ▶ Chap. 1. The major processability factors for phenolic resins are as follows:
   - The viscosities of phenolic resins are low.
   - Phenol formaldehyde resin cures at high temperature (such as 150–160 °C) and does not require any curing agent or hardener.
   - During curing, water of the reaction needs to be removed from the composite part.
   - Phenolic resins have problems advancing in the reaction; therefore, the formation of tiny cross-linked lumps of resin causes difficulty with its processability.
   - The resin must be stored under cool and dry conditions. The advancement of reaction leading some cross-linking may be identified by the presence of water (water of reaction) on the upper surface of the resin.
   - Phenolic resins have low shrinkage (<0.5%).
5. *Bismaleimide Resins*—Bismaleimide (BMI) resins are high-temperature resins and are abundantly used in the composite industry. They are used in applications where good thermal performance coupled with good mechanical properties is needed. Their mechanical performance is inferior to epoxy and isophthalic UPE resins. BMI resins are prepared from a condensation reaction of maleic anhydride and aromatic diamine (such as DDM). Water from the reaction comes out as

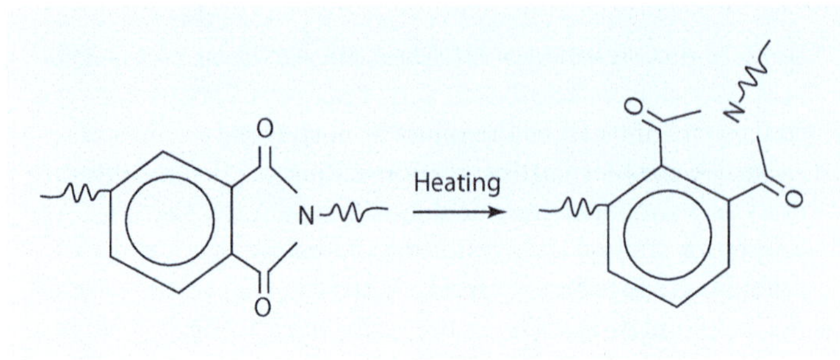

Fig. 5.10 Energy absorption by an imide linkage and its shifting position mentioned in ▶ Chap. 1.

a condensate. BMI resins are prepared in two steps: first a polyamic acid is formed and second a further condensation reaction makes a bismaleimide prepolymer resin. This is cured in the presence of an amine (such as DDM or DETA). Amines are premixed and are used as one component system. The BMI resin shelf life is short, and it must be stored under cool and dry conditions. The mechanical properties of bismaleimide resins have been mentioned in ▶ Chap. 1. The major processability factors for epoxy resins are the following:
- The viscosities of bismaleimide resins are low, and it is easy to process them.
- They are cured at a high temperature (190–230 °C).
- Their T_g is in the range of 190–230 °C.
- Bismaleimide resins have low shrinkage (> 0.5%).

6. *Polymeric Monomer Reactant-Polyimide (PMR–PI) Resins*—PMR–PI resins are the highest-temperature resins known to date. Their service temperature is up to 367 °C. PMR–PI resins are thermally stable at high temperatures due to the shifting of their carboxylic linkage. At high temperatures, an imide linkage keeps on absorbing energy and dissipates that energy by shifting its position on the benzene ring, as shown in Fig. 5.10. PMR–PI resins are important for supersonic and hypersonic aerospace applications. The mechanical properties of PMR–PI resins are similar to bismaleimide, as
   - The viscosities of PMR–PI resins are low and easily processable using VARTM and autoclaving methods.
   - PMR–PI resins cure at high temperatures (such as 325–367 °C) and do not require any curing agent or hardener.
   - Their T_g is also in the range of 325–367 °C.
   - During curing, condensates (such as methanol/ethanol) and solvents (such as dimethylacetamide/N-methylpyrrolidone) need to be removed from the composite part.
   - PMR–PI resins are used in a vacuum-assisted process to remove high boiling condensates.
7. *Polyurethane (PU) Matrix Resins*—PU resin is made from an addition reaction of polyol and polyisocynate. This resin provides a good surface finish and can be used for VARTM and high-pressure RTM (HP-RTM). It is used as a matrix with

carbon and glass fibres. Polyurethane matrix composites have excellent strength coupled with toughness. PU matrix resins can be used for automotive exteriors and sports goods (such as tennis racquets, squash racquets, etc.) The general properties of polyurethane matrix resins have been mentioned in ▶ Chap. 1. The major processability factors for polyurethane matrix resins are the following:
   - Polyurethane matrix resins are low-viscosity resins and cure at a moderately low temperature (T_g = 120 °C). Thus, they do not cause internal or residual stresses in the component.
   - PU matrix resins may be demoulded at high temperatures due to its thermal stability. Therefore, it enhances the demoulding rate and thereby improves the rate of production.
   - PU matrix resins are susceptible to UV degradation; therefore, they need UV stabilizers in their formulations.
   - The inherently good surface finish and reactivity of PU matrix resins facilitate printing on the finished composites.
   - These resins have an extraordinarily good interphase with carbon fibres due to the reactivity of PU with the sizing of carbon fibres.
   - PU matrix resins have low shrinkage (>0.2%).
8. *Novolac Resins*—Novolac resins are commonly known as novolacs and are generally available in two types: phenolic and epoxy. Novolacs are cured by an addition reaction and do not release any condensate. They are used in the form of prepregs, as described earlier in this chapter. Phenolic novolacs are mostly used in the form of glass fibre CSM prepregs and are processed using compression moulding. The curing temperature of both phenolic and epoxy novolacs are 160 °C and 180 °C, respectively. The general properties of novolacs are similar to the properties of the parent resins (such as phenolic or epoxies), as mentioned in ▶ Chap. 1. The major processability factors for novolac resins are given here.
   - Novolacs are made in powder form. Initially, they melt on heating, and then subsequently—up to their curing temperature—they get cross-linked and cured.
   - They need post curing in a closed mould at 120 °C to remove residual stresses in the composite product.
   - Novolacs have low shrinkage (>0.5%).

5.3.4 Selection Criteria for Reinforcement

Reinforcements, as we know, are the main building blocks of a composite as they are the load-bearing members and are responsible for most of the major properties of composites [18]. Their properties, size, and shape are important for processability. The processability of reinforcements and matrices vis-à-vis processing methods is shown in ◘ Table 5.2. The selection criteria for the reinforcement of

Table 5.2 Processability of reinforcement and matrices vis-à-vis processing methods

Serial No.	Reinforcement	Processing Method	Matrices
1.	Synthetic fibres (glass, carbon, and aramid fibre)		All thermosetting matrices
	(a) Short fibres/ staples	Hand layup and compression moulding	
	(b) Long fibres	All moulding processes except injection moulding	
2.	Natural fibres		All thermosetting matrices except high-temperature curing matrices, i.e. PMR–PI
	(a) Short fibres/ staples	Hand layup and compression moulding	
	(b) Long fibres	All moulding processes except injection moulding	

thermosetting composites are dependent on the processability factors of the reinforcement as well as the processing technology [19–21]. The processability factors for reinforcement are detailed here.

1. *Properties Driven by Selection Criteria*
   - *Form of Reinforcement*—In thermosetting composites, all forms and shapes of reinforcements, as described in earlier sections, are used. There are specific restrictions for the use of any weave, staple, particles, etc. of the reinforcement for the fabrication of composites.
   - *Wettability of Reinforcement*—The surface of the reinforcement must have a high surface energy to allow the resin to wet its surface. The wettability of the reinforcement will ensure a good interphase between the reinforcement and the matrix. This will result in a good composite.
   - *Chemical Interaction with Resin*—The hydrophobic surface of the reinforcement will allow the formation of a chemical bond with an organic matrix. This will yield into a good interphase. Reinforcement must be selected based on its capability of having a good chemical interaction with the matrix.
   - *Fire Resistance*—This is an important parameter from the point of view of safety. Most synthetic-reinforcing fibres are fire safe except UHMWPE, PP, and PEEK. Natural fibres are susceptible to fire hazards, and care must be taken that

the processing temperature is less than their ignition temperature.

- *Thermal and Hygrothermal Stability*—The thermal stability of synthetic fibres (such as glass, carbon, and Kevlar fibres) is good, and they can be used under any thermal processing. Natural fibres are thermally less stable, and care must be taken for their processing as mentioned previously. Moisture affects all fibres, and the ingress of moisture in Kevlar deteriorates its mechanical properties. In glass fibres, moisture causes depletion of the coupling agent. In carbon fibres, moisture affects the polyimide or epoxy over the polymeric sizing. Natural fibres absorb moisture; thus, their mechanical properties are diminished.
- *Post-processing Operations*—The role of reinforcement and its machinability is important for post-fabrication operations. Kevlar-reinforced composites are difficult to machine, and they require water jet or laser cutting operations for smooth cutting edges. Kevlar, having a fibril structure, causes the fibrillization of fibres and gives a rough surface to mechanical machining using standard cutting tools, such as a hacksaw, milling cutter, turning tool, etc. The processability of composites does not end with striking the correct combination of raw materials and processing methods to just get the product moulded. The designer and process engineer must be prudent enough to also plan the processing of the composite products for post-fabrication finishing. Thus, the raw materials, moulding surfaces, and processing technologies need to be chosen in such a manner that they ensure a minimum of post-fabrication operations on moulded composite products.
- *UV Resistance*—UV radiation does not affect natural fibres and most synthetic-reinforcing fibres (except polyolefin fibres, such as PP, UHMWPE, etc.).

2. *Processing Technology Driven by Selection Criteria*
   - *Compatibility with the Form of Reinforcement*—The forms and weave patterns of reinforcement have a direct effect on the processability of composites. Fabrics, CSM, and staples cannot be used for filament winding and tape winding processes. Staples cannot be used for the VARTM process. Natural vegetation fibres are not compatible with high-temperature processing, and they also have limitations for a filament winding process due to their limited tensile strength.
   - *Flow of Matrix*—The weave of a reinforcement must allow the resin to flow through it so that complete wettability is achieved.
   - *Production Rate*—The production rate of composites also has a bearing on the type of reinforcement to be used. In order to achieve high production rates, prepregs or performs are preferred to be used. For fewer batches (e.g. ten pieces) of composite products, the hand layup method may be the best in order to minimize mould costs, leading

5

to lower production costs. Whereas if the same product is required in larger batches, resorting to compression moulding technology with a matched die mould is the best option. Then the cost of the tooling could be amortised faster with large production volumes. For higher production volumes, the processability of raw materials in loose forms (e.g. reinforcing fabric and matrix) with moulding processes such as matched die moulding on compression moulding is not suitable. Rather, the use of sheet moulding compounds (SMCs) or bulk moulding compounds (BMCs) offers better processability over the loose form of raw materials to match the production volume and quality.

- *Cost of Production*—The cost of production may be controlled by choosing the right kind of reinforcement for a required service and the service life of the composite. Also, the form and weave of the reinforcement could impact the cost of the product. If CSM can provide the necessary performance of the composite, using woven fabric would unnecessarily raise the cost.

5.4 Methods to Improve Processability in Thermosetting Composites

Point to Ponder…

***A** big crackling sound is heard when a glass vase is broken. The energy released in the breaking of the bonds is heard by us in the form of sound (mechanical energy). This energy is the work required to separate the bonding between glass molecules of the vase.*

The properties of composites depend upon the simultaneous and cooperative interaction between the matrix and the reinforcement. The following are the basic requirements to cause synergistic properties of composites.

5.4.1 Adsorption and Wetting

The wetting of fibres and the matrix is essential to form a continuous interphase region. According to the surface chemistry, the work (energy) required to separate two surfaces can be defined in terms of the surface free energies of two particles, A and B, as

$$W_{\text{adhesion}} = \gamma_{\text{A}} + \gamma_{\text{B}} - \gamma_{\text{AB}} \quad (5.1)$$

where

- W_{adhesion} = Work of adhesion
- γ_{A} = Surface free energy of particle A
- γ_{B} = Surface free energy of particle B
- γ_{AB} = Surface free energy of interaction between particles
- A and B

The summation of the work of adhesion of all the particles on the surface gives the total work of the adhesion of the surface. The surface free energy can be approximated by making surface tension measurements. However, the work of adhesion should not be equated to the bond strength, since the bond strength also contains the energy of dissipative processes (such as viscoelastic deformation, microcracking, plastic deformation, etc.) However, under certain conditions (low temperature and high peel rate), the work of adhesion ($W_{adhesion}$) approximates the bond strength (W_{bond}).

Surface adhesion is essentially a surface energy phenomenon and can be enhanced by altering the surface energies of the fibre and/or the matrix.

5.4.2 Chemical Bonding

Chemical bonding between the matrix and the reinforcement is the most effective method of increasing the interfacial bond strength. Some fibres have chemically active groups that react with the matrix, and in some cases, coupling agents are used to enhance the bonding. In the case of inorganic reinforcements (such as glass fibres, silica fibres, etc.), silane or titanate coupling agents are used, as discussed in previous sections. For thermosetting matrices, silane coupling agents are preferred, and for thermoplastic matrices, titanate coupling agents are preferred. Plasma is also used to provide improvement in interfacial adhesion for chemically inert materials (such as polypropylene, polyethylene, and Teflon). Plasma extracts atoms from the fibre surface and promotes the attachment of reactive groups, which can bond with appropriate chemical group of the matrix. The use of a corona discharge has also shown some improvement in bonding between matrices and chemically inert fibres [22].

5.4.3 Mechanical Adhesion

Mechanical adhesion takes place due to irregularities on the surface of the reinforcement. These irregularities could be inherent but also could be introduced on the surface of the reinforcement. Surface irregularities are shown schematically in ◘ Fig. 5.11. In the case of aramid fibres (such as Kevlar), the surface is very smooth due the

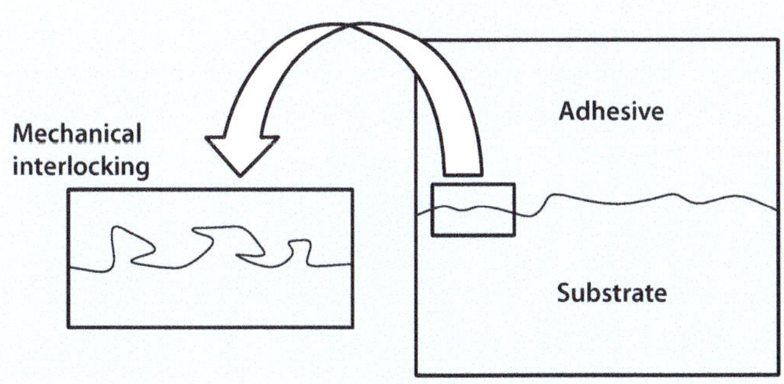

◘ **Fig. 5.11** Schematic representation of mechanical locking

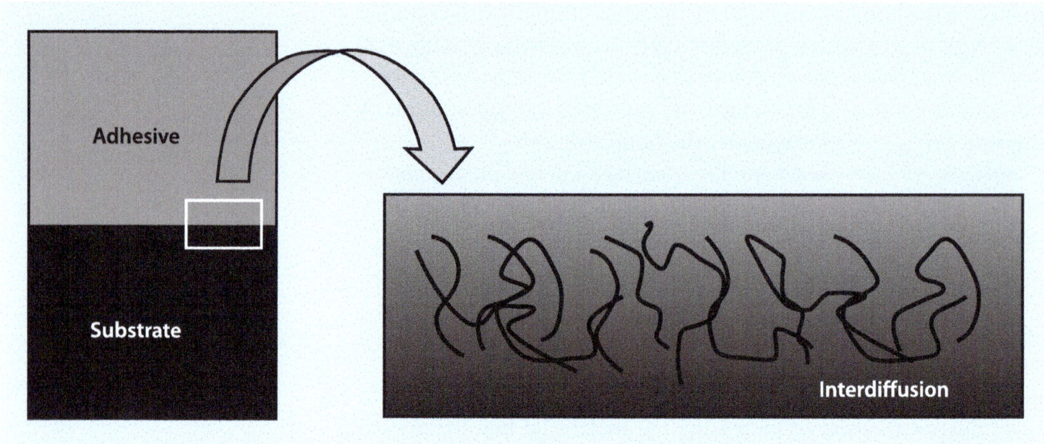

Fig. 5.12 Schematics of interdiffusion

highly regular and oriented molecular structure of aramid fibres. Though the chemical bond with the matrix could be formed due to the chemical nature of aramid, wettability remains the problem to achieving a good chemical bond. Therefore, the surfaces of aramid fibres are mechanically roughened using quartz stone. This rubbing against the quartz stone causes the microfibrillization of aramid fibres on their surface, and this leads to good wettability, mechanical bonding, and chemical bonding due to a higher surface area. Boron carbide fibres and silicon carbide fibres are made by depositing a carbide layer over tungsten or a carbon filament at a high temperature and in an inert atmosphere. These fibres are high-performance fibres having a high modulus and with a high-temperature resistance. The rough surface created by depositing the carbide provides the location for mechanical interlocking of the matrix.

5.4.4 Interdiffusion

The mutual solubility of two materials is the first approximation to the amount of interdiffusion. The molecular intertwining can increase the strength of the adhesive bond. The interdiffusion is shown schematically in Fig. 5.12. High crystalline polymers have relatively poor interdiffusion below their T_g. The property of interdiffusion in composites causes the adverse effect on reduction of fibre dimension, therefore, reduction in mechanical properties of composites. Also, high solubilities may lead to a solution rather than having a composite where the constituent phases should be distinct. The interdiffusion process of adhesion has found positive results in porous reinforcement, which strengthens by the diffusion of the matrix.

5.4.5 Electrostatic Attraction

Adhesion between fibre and resin may be promoted, at least initially, by a difference in the electrostatic charge between two materials. For high-performance composites and some of the engineering thermoplastics, the mechanisms of adhesion and matrix-reinforcement bonding are poor. Adhesion between thermoplastic resins and fibres is usually weak because of poor adsorption and wetting, the absence of chemical bonding, a lack of mechanical adhesion, limited or no interdiffusion, and poor electrostatic attraction. The surface free energies, γ, for thermoplastics are usually much lower than those for conventional thermosetting resins, resulting in a low $W_{adhesion}$. Wetting of fibres with thermoplastics is also poor because of their high viscosities. High viscosity, low solubility, and high crystallinity of thermoplastics also limit the interpenetration of the interphase regions. Thermoplastic resins have few active sites—especially in highly aromatic resins—that can form chemical bonds with the surface of the reinforcement. Mechanical bonding does not take place for smooth surfaces and reinforcements free of cracks. Also, highly viscous thermoplastic resins flowing into small pockets or crevices on the surface of the reinforcement make it difficult to create good mechanical bonding. Generally, no electrostatic bonding is present in thermoplastic composites [23–25].

In order to achieve good bonding between thermoplastic matrices and fibre reinforcements, several techniques providing a conventional layer to reduce interfacial stress concentration are used. Example of such techniques are the electrodeposition of a polymer on the fibre surface, improvement in the coupling agents, improved surface oxidation treatments, matching mechanical properties (especially the modulus) of the outer surface of the fibre and matrix to lower the stress concentration that occurs at the surface during loading, and using a ductile matrix in the interphase. Matching the modulus of the outer surface of the fibre and the matrix leads to graded reinforcements throughout the cross section.

It is a fact that in composites, the properties of reinforcement are far, far superior to matrices in most cases. According to the rule of mixtures, the overall properties of composites are compromised due to this fact. The reasons for these facts lay in the following:

- Selecting or developing a matrix that matches the high properties of the reinforcement
- Improvement of the interphase
- Removal of surface and core defects in injection-moulded composites
- Achieving perfection in the desired orientation of the reinforcement

New developments have suggested better matrices for the composites—especially for carbon fibre-reinforced composites where epoxy resins are primarily used. Higher functioning epoxy resins have been developed to improve the properties of the resultant composites. These epoxy resins are tri-functional and tetra-functional resins. A new variety of polyester resins, such as vinyl ester resins, has changed the scenario of polyester-based composites. For high-temperature applications, polyimides (PI) have a distinct place. A range of polyimides (such as bismaleimide, polymeric monomer reactant–polyimide (PMR–PI), polyetherimide, polyamide imide, etc.) are being used commercially. Nevertheless, the future of composite technology will greatly depend upon the development of new technologies and the improvement of reinforcements, matrices, and interphases. Future technologies need research and development efforts for their commercialization, especially the crystallization and growth of polymeric matrices over a reinforcing fibre, selective irradiation methods for creating a fibre structure in a neat matrix in situ, automatic swelling of the matrix to minimize the air pockets and gaps in the composites, etc. Chemically active reinforcements are responsible for the strong interphase in present composites. If the surface is inert or less active, it needs to be made chemically active, as described in the previous section. However, the chemical activity needs further development and characterization for inducing suitable and compatible linkages (such as dipole interactions, hydrogen bonds, van der Waals forces, ionic interactions, etc.) in order to optimize the bond strengths along with its lability in the interphase [26].

Processing technologies (such as autoclaving, RTM, filament winding, etc.) provide composites almost free of defects. Similarly, filament winding technology, tape winding technology, etc. are used for a high orientation of reinforcement.

5.5 Processability of Thermosetting 'Green Composites'

Point to Ponder…
Bulk density is the mass of aggregate of material (in the form of bunch, bundle, powder, etc.) per unit volume. It is not intrinsic property of a material.

Green composites are those which do not cause any harm to the environment during both their processing and the product disposal after the useful life is over. Given this classification, major composites are those that use natural fibres as reinforcements and/or biodegradable matrices. Green composites have created interest worldwide due to their environment friendliness. These composites are not new. In fact, mankind's knowledge of green composites has existed from time immemorial. In ancient times, straw- or bamboo-reinforced mud houses were built (as mentioned in ▶ Chap. 1). More recently, natural fibres (such as sunn, coir, jute, etc.) have been used as reinforcement in the construction of buildings. The densities of these fibres are in the range of 1.15–1.55 g/cm^3, which supports their processability using common processes and the development of matrices that are processable within the limit of thermal degradation of natural fibres. Rubberized cotton, which historically has been used in bicycle tyres, is an apt example of a natural fibre-reinforced polymeric composite.

Example 5.3 Cast some light on green composites with their specific applications

Answer

During World War II, flax fibre-reinforced phenolic resin was used to manufacture fuselages of fighter aircraft. Natural fibres were first used by Henry Ford in automobiles when he used hemp fibre-reinforced composites in 1942. Presently, most German cars use natural fibre-reinforced composites for boot linings, back door panels, back seat fittings, dash boards, etc. Natural fibre composites are mostly used in the automotive industry. Presently, BMW cars use 8–13 Kg natural fibre-reinforced composites per vehicle, and Ford cars use 5–13 Kg natural fibre-reinforced composites per vehicle.

The use of biodegradable fillers (such as starch, wood flour, etc.) in conventional thermoplastic or thermosetting matrices increases the scope of biodegradable matrices. It is to be noted that the bulk of natural fibres is quite high, and this makes their weight percentage in the composite quite low. The processability of thermosetting green composites is governed by the following factors:

- *Interphase in the Composite*—The hydrophilic nature of natural fibres restricts their wetting by hydrophobic polymeric matrices. This may limit the bond strength, resulting in poor composite properties. In order to make natural fibres processable with polymeric matrices, it is essential that they are chemically treated in order to attach chemical groups on the fibre surface as potential sites for bonding with the polymeric matrix. Generally, natural fibres are treated with weak acids – (—COOH, —CO, —CHO, etc.) to make attachments onto their surface.
- *Hygroscopic Nature of Natural Fibres*—The hygroscopic nature of natural fibres may lead to imperfections of the morphology after moulding or to dimensional inaccuracies of the composite being developed. Adequate drying of the natural fibres is needed before processing their composites. The ingress of moisture in the composite also will reduce its properties.
- *Low Strength*—Low strength of natural fibres restricts their processability using either filament winding, tape layup, or pultrusion methods.
- *Use of High-Temperature Matrices*—Due to the low thermal stability of natural fibres, their processability as reinforcement in high-temperature matrices (such as PEI, PEEK, polyimides, etc.) is not feasible.
- *Service Environments*—Service environments having high temperatures or a high moisture content are not suitable for natural fibre-reinforced polymeric composites.
- *Safety*—The processability of these composites does not cause any health hazard because the natural fibres are not toxic.

5.6 Processability of Smart and Intelligent Composites

Emphasis is being given on the research and development of smart and intelligent composites. It is easy to develop smart and intelligent structures in composites due to tailorability of their properties and possibility of hybrid structures in them. The processability of smart and intelligent composite structures involves packaging of necessary sensors, mechanisms for obtaining desired output, and healing response to that output [27].

5.6.1 Processability of Smart Composites

Smart composites or structures are those which can communicate their status or condition to the observer or designer. Invariably, these structures have mounted or embedded sensors that display the healthy properties of the composite. For example, if a multilayer composite is manufactured with different coloured laminas, erosion of the uppermost layer will reveal the exposed layer to be a new colour. This new colour will indicate the number of eroded layers in the structure because with every layer eroded, a lamina with a different colour will appear. Similarly, a strain sensor is embedded at the location prone to failure. This sensor provides information about the level of strains and failure due to excessive strain. The processability of smart composites is similar to the processability of other advanced composites; however, some other factors are outlined here.

- The thermal stability of embedded sensors should be high enough to sustain the curing temperature of the thermosetting resin.
- Sensors must be able to sustain the moulding pressure—if any.
- The antenna or communication mechanisms must not weaken the composite structure.
- Arrangement for the power source of the sensor to remain activated for the lifespan of the composite structure must be addressed.

5.6.2 Processability of Intelligent Composites

Intelligent composites or structures are those which can identify the defect and identify it. They have mounted or imbedded sensor(s) along with materials and mechanisms to repair the composite at the defect site. Therefore, intelligent composites are able to repair their defects without external help. For example, epoxy resin-filled glass microbeads and hardener-filled glass microbeads are placed in the intelligent composite during its processing. Under excessive strains, composites may experience delamination or internal cracks. Microbeads are designed to break under excessive strain so the resin and hardener will ooze out, filling the crack and curing at room temperature. That's how a crack would be repaired.

Sensor-based intelligent composites have sensor(s) mounted or embedded in the composite along with the repair material and actuator. For example, intelligent composite bridges have embedded and/or mounted sensors to sense the development of a crack. A signal is received by the actuator, and it releases the requisite amount of repair material at the crack site automatically. As a result, the crack is filled and is cured at room temperature. The only requirement in such cases is to check the quantity of the repair material in the reservoir so that enough repair material is available. The processability of intelligent composites is similar to that of other advanced composites; however, some other factors are outlined here.

- Thermal stability of the embedded sensor and repair mechanism should be high enough to sustain the curing temperature of the thermosetting resin.
- Sensor and repair mechanisms must be able to sustain the moulding pressure—if any.
- The sensor and repair mechanism must not weaken the composite structure.
- The shelf life of the embedded repair material must be long enough to sustain the serviceability and repair of the composite structure.
- Mechanisms used to replenish external repair material must be available.
- Arrangement for the power source of the sensor to remain activated for the lifespan of the composite structure must be addressed.

To Be Remembered

1. *Interface* is the bonding surface between the reinforcement and the matrix across which a discontinuity in some parameter occurs.
2. *Interphase* is the continuous region of dissimilar composition between the reinforcement and the matrix.
3. There is a labile link between the reinforcement and the matrix through interphase. This labile link causes the shifting of the interphase bond position.
4. *Particles* are regular-shaped, small-sized entities.
5. *Whiskers* are irregular-shaped, small-sized entities.
6. *Cure cycle* is a time and temperature combination used to cross-link the thermosetting resin, providing optimum properties.
7. *Cross-link density* is the direct measure of the extent of curing for thermosetting resins.
8. *Wetting* the fibres of the matrix is essential to form a continuous interphase region.
9. *Chemical bonding* between the matrix and the reinforcement is the most effective method of increasing the interfacial bond strength.
10. The mutual solubility of two materials is interdiffusion.

5

Exercises

1. Predict and draw the curing cycle of a thermosetting resin having a gelation temperature and curing temperature of 70 °C and 160 °C, respectively. It has an initial viscosity and viscosity at 70 °C of 12,000 cps and 2000 cps, respectively.
2. Compare the fibre treatments on glass fibres, carbon fibres, and Kevlar fibres. How is the role of sizing different in glass fibres and carbon fibres?
3. What are the polyolefin fibres? Why are polyolefin fibres not used in the fabrication of high-performance composites?
4. A composite and steel adhesive joint has a lap shear strength of 2.2 Kg/mm^2. The surface free energy of bonding the steel and composite is 0.7 J and 0.9 J, respectively. Calculate the interaction energy of the composite and steel joint in Joules.
5. How is interdiffusion different from mechanical locking?
6. What are the factors that influence the following in composites?
   (a) Mechanical locking
   (b) Chemical bonding
7. Why is the role of secondary chemical bonding important in the bonding of fibres and matrices? What will happen if there is only primary bonding?
8. Explain thermodynamic stability in cases of chemical bonding and mechanical locking in the processing of polymeric composites.
9. What are the reasons for the popularity of prepregs in composite industries despite that the fabrication of composites using prepregs is a two-step process causing a high cost of production?
10. How does the reinforcement from particles and whiskers in a matrix differ in the performance of polymeric composites? Which is better and why?
11. In thermoset composites, the temperature of the onset of the gelation of a resin is 90 °C, and the temperature of the onset of degradation of a resin is 270 °C. What is the likely curing temperature of this composite and why?

References

1. Hon, D. and Shiraishi, N., Eds., *Wood and Cellulose Chemistry*, 2nd Edition, Marcel Dekker, New York, 2001.
2. Shaffer, G.D., "An Archaeomagnetic Study of a Wattle and Daub Building Collapse," *Journal of Field Archaeology*, 20 (1) (1993), p. 59.
3. Pukanszky, B., Tudos, F., Kolarik, J., and Lednicky F., "Ternary Composites of Polypropylene, Elastomer, and Filler: Analysis of Phase Structure Formation", *Poly. Compos.*,11 (1990), p. 98.
4. Brydson, J.A., *Plastics Materials*, Sixth Edition, Butterworth Heinemann, USA, 1995.
5. Guth, E. and Smallwood, H., "Limiting Law of the Reinforcement of the Rubber", *J. Appl. Phy.*, 15 (1944), p. 758.
6. Pludeman, E., Collins, N., *Adhesion Science and Technology*, Vol. 9a, Plenum Press, New York, 1975.
7. Monte, S.J., *Ken-React Reference Manual- Titanate, zirconate and aluminate coupling agents*, 1985.

8. Hopgarten, H., "Surface study of carbon fibres with esca and auger electron spectroscopy", *Fibre Science and Technology*, 11 (1978), p. 67.
9. *Carbon Fibres*, Wikipedia, https://en.wikipedia.org/wiki/Carbon_fibers.
10. "Natural Wood Fibres to Reinforce Plastic Materials," *FRP Today*, 14(12), December 2014, p. 23.
11. *Fibreglast, 6K, 2X2 twill weave carbon fibre fabric*, www.fibreglast.com.
12. "Nobel Natural Fibre Composite Solution for the Automotive," *FRP Today*, 15(5), May 2015, p. 21.
13. Varma, I.K., Sangita, and Varma, D.S., "Addition Polyimides. IV. Effect of Structure on Thermal Characteristics", *J. Poly Sci.*, 29 (1984), p. 2807.
14. Siegmann, A. and Turi, E.A., J. Macromol., "Structural Changes in Glassy Poly(Ethylene Terephthalate)", *Sci.- Phys.*, B10(4) (1974), p. 689.
15. Lakkad, S.C. and Varma, I.K., eds., "Design and Evaluation of Some FRP Products, Seminar Proceedings- Matrix Resins for Composites," *Department of Science and Technology*, 1986, p. 420.
16. "New Technique to Heat and Cure Composite Materials," FRP *Today*, 15(5), May 2015, p. 24.
17. "Ultra Durable CFRP High Pressure Hydrogen Fuel Storage Tank," Maruhachi Co. Ltd., Japan, *FRP Today*, 14 (3), March 2014, p. 23.
18. Corten, H.T., *Engineering Design for Plastics*, Baer E., Ed., Reinhold Publishing Corporation, New York, 1964.
19. "Post Curing of Resins," *FRP Today*, 14(4), April 2014, p. 8.
20. Young, R., *Handbook of Fibre Glass and Advanced Plastics Composites*, Lubin, G., ed., Van Nostrand Reinhold Company, New York, 1969.
21. Gandhi, K.S. and Burns, R., *"Rheological Properties of Glass-reinforced Dough Moulding Compounds"*, *Trans. Rheo Soc.*, 20 (1976), p. 489.
22. Hayes, B.J., "Prepreg Technology, Polymers and Composites-Recent Trends," *Department of Science and Technology*, 1989, p. 345.
23. Suddel, B.C. and Evans, W.J., *Natural Fibre Composites in Automotive Applications, Natural Fibres, Biopolymers and Biocomposites*, Mohanty A.K., Misra M. and Drzal L.T., Ed., CRC Press, Taylor & Francis, 2005, Chapter 7.
24. Sabir, M.I., Xu, X., and Li, L., "A review of biodegradable polymeric materials for bone tissue engineering applications," *J. Mater. Sci.*, 44(21) (2009), p. 5713.
25. Taylor, A., "Case study in fibres in composite materials e.g. hemp in automotive applications," *Fourth Meeting of Government Industry Forum on Non-food uses of Crops, GIFNFC4/4 Fibres in Composite Materials*, DTI Conference Centre, London (cited as ref. 30 in ref 16 of this chapter), 22 January 2002.
26. "Loctite MAX 3- the Latest Generation of Polyurethane Matrix Resins," *FRP Today News*, 13 (10), October 2013, p. 26.
27. Kempf, M., Schwaegele, S., Ferencz, A., and Altstaedt, V., "Effect of impact damage on the compression fatigue performance of glass and carbon fibre reinforced composites," *18th International Conference on Composite Materials*, Jeju, Korea, 2011.

Processability of Thermoplastic Composites

A. K. Ghosh, M. Dwivedi, *Processability of Polymeric Composites*,
https://doi.org/10.1007/978-81-322-3933-8_6

The basic principles of processability and the factors affecting it do not change much from thermosetting composites to thermoplastic composites. However, the processability *factors*, which have a profound effect on the processing of thermoplastic composites, are mentioned here.

- Thermoplastic matrices are generally solid to start with.
- Thermoplastic composites are not processed at room temperature but only at high temperatures.
- Thermoplastic composites have melt viscosities that are higher than those compared to thermosetting matrices.
- Thermoplastic composites are categorized as crystalline, such as nylon; semi-crystalline, such as polyether ether ketone (PEEK), polyethylene (PE), and polypropylene (PP); and amorphous, such as polyether imide (PEI), polymethyl methacrylate (PMMA), and polycarbonate (PC) in nature.

Thermoplastic composites are easily recycled; therefore, they are finding many applications in composite industries [1, 2]. They are easily blendable, and no additive or hardener (like in the case of thermosetting matrices) is required to solidify or cure the thermoplastic composites. The ability of thermoplastic composites to repair through welding is also an advantage. However, the downside is that its mould designs are complex and heavy processing machines are required.

This chapter delves in detail for better insight on the processability of thermoplastic composites.

6.1 Interface and Interphase in Thermoplastic Composites

The concepts of interphase and interface are discussed in ► Chap. 5 and remain same for thermoplastic composites. Olefinic matrices are often inert. Polypropylene (PP), low-density polyethylene (LDPE), high-density polyethylene (HDPE), etc. are chemically inert, and they each can form a bond with the reinforcement with great difficulty. Special treatments on fibres are carried out in order to improve the bonding between the reinforcement and the thermoplastic matrix. Thermoplastic fibres (PEEK, nylon, polyester, polyurethane (PU), etc.) are chemically active and make a good interfacial bond with reinforcements. Their fibres are also chemically active—except for ultra-high molecular weight polyethylene (UHMWPE) fibre. UHMWPE fibres are not treated with any chemical and are transformed into composites that have low-temperature matrices that are compatible with polyethylene. The matrices used for UHMWPE fibres are low-temperature elastomers and polyolefins (polyethylene and polypropylene). The interfacial bond made by these fibres is a secondary bond, such as found in Van der Waals interactions [3, 4].

6

6.2 Forms of Reinforcements for Thermoplastic Composites

The most common processes used for the fabrication of thermoplastic composites are injection moulding and compression moulding methods. Short fibres, staples, particles, whiskers, etc. are used commonly as reinforcement for thermoplastic matrices in the injection moulding process. A chopped strand mat (CSM) is not used in thermoplastic composites because the wettability of fibres is difficult and can increase the cost of production. Natural fibres are used with care so that their thermal stability is more than the processing temperature of the composite. Reinforcement in the form of fabrics is used with compression moulding. Prepregs (fabric/rovings/tows) are also used for specialized applications, such as hollow weight-bearing structures, pressure vessels, etc. [5–7]. The forms of reinforcement used in thermoplastic composites are the same as those mentioned in ▶ Chap. 4.

6.3 Models for Mechanical Properties in Particulate Thermoplastic Composites

Various properties also can be calculated from idealized models that closely approximate real composite behaviour. Models applicable to reinforcing systems for tensile strengths and tensile modulus are given in this section.

6.3.1 Models Applicable to Tensile Strength

Pukanszky developed a simple model to describe the composition dependence of the tensile yield stress of particulate filled composites. This model is based on the fact that an interphase forms spontaneously in composites, and it is assumed that yield stress changes proportionally to its actual value as a function of composition [8–12]. Accordingly, the composition dependence of tensile yield stress can be described as

$$\sigma_c = \frac{\sigma_m (1-\phi_f)}{1+2.5\phi_f} \exp\left(B_y \phi_f\right) \tag{6.1}$$

where σ_c and σ_m are the yield stress of the composite and the matrix, respectively. ϕ_f is the volume fraction of the filler in the composite. The following term expresses the effective *load-bearing cross section index* (BI) of the matrix:

$$\frac{1-\phi_f}{1+2.5\phi_f} \tag{6.2}$$

The effective load-bearing cross section is a factor. At zero interaction, all of the load is carried by the polymer, and the load-bearing cross section decreases with increasing filler content. The same correlation can be used to describe the composition dependence of the tensile strength if the elongation of the composite is small, usually less than 100%. A detailed study of numerous composites has proven that, in composites containing stiff fillers or reinforcements, a hard interphase is formed. This increases the load-bearing capacity of the filler and contributes to the reinforcement. Only the formation of such a hard interphase explains that yield stresses of a composite exceeding the yield stress of the matrix are achieved occasionally. The load carried by the second component depends also on the properties of the matrix; the extent of reinforcement is larger in a softer than in a stiffer polymer. This factor also must be taken into account when composites prepared with different matrices are compared with each other.

This model was developed specifically for composites containing spherical particles. The BI values reported for various systems are 9.7 for PBT/ABAS/Mica ($\phi_f < 0.07$) and 10.4 for TiO_2-filled PBT/ABAS blends ($\phi_f < 0.25$). PP filled with $CaCO_3$ showed a BI value of 1.5, and montmorillonite (MMT) clay-filled PP showed a BI value of 1.8. On the other hand, a MMT-filled PP with a fully exfoliated state showed a BI value of 190–200.

Example 6.1 Calculate effective load-bearing cross section index of a particulate composites having ϕ_m 0.7.

Answer

$$\phi_m = 0.7$$

$$\phi_f = 0.3$$

So, the effective load-bearing cross section is

$$\frac{1-\phi_f}{1+2.5\phi_f} = 0.35$$

6.3.2 Models Applicable to the Tensile Modulus

Kerner predicted a model by taking Poisson's ratio of the polymer for rigid particles composites as

$$E_c / E_p = 1 + \left[\frac{15(1-\nu_p)}{8-10\nu_p}\right]\left(\frac{\phi_f}{1-\phi_f}\right) \tag{6.3}$$

where

- E_c = Modulus of the composite
- E_p = Modulus of the polymer
- ν_p = Poisson's ratio
- ϕ_f = Fraction of the filler

Guth and Smallwood also proposed a relation to predict the modulus of particulate filled composites by

$$E_c / E_p = 1 + 2.5\phi_f + 14.1\phi_f \quad \textbf{(6.4)}$$

where ϕ_m is the volume fraction of the matrix.

Example 6.2 Calculate the modulus of alumina whiskers in reinforced PP where alumina is 30% by weight. The specific gravities of PP and alumina are 0.94 and 2.6, respectively. The modulus of PP is 3 GPa.

6

Answer

Volume of alumina = Mass/Density
Volume % of alumina = 30/2.6 = 11.54%
Volume % of PP = 88.46%
$\phi_r = 0.1$
$\phi_m = 0.9$
$E_c/E_p = 1 + 2.5\ \phi_r + 14.1\ \phi_m$
So,
$E_c = 41.8$ GPa

6.4 Thermoplastic Composites

Thermoplastic composites have created a niche for themselves due to their ease of processability, repeated quality, good damage tolerance, and low risk of environmental hazards. They are finding applications in automobiles, aviation, packaging, refrigeration, etc. [13–16]. The moulding of thermoplastic composites does not involve any curing mechanism; rather, it involves the melting of the matrix and its solidification as a function of heat. The characteristics of thermoplastic matrices are given in Table 6.1.

Table 6.1 Characteristics of thermoplastic matrices

Serial No.	Parameters	Thermoplastics
1.	Shelf life	Unlimited
2.	Fabrication time	Short
3.	Post formability	Possible
4.	Generation of by products	Almost nil
5.	Tooling	Expensive
6.	Mixing of chemicals	Not required
7.	Production rate	Fast
8.	Economy	High
9.	Handling	Easy
10.	Recycling	Possible

Example 6.3 Calculate Poisson's ratio of a talc-filled HDPE composite. The modulus and ϕ_f of talc is 12 GPa and 0.2, respectively.

Answer

$$E_c = E_p(1+2.5\ \phi_f + 14.1\ \phi_f)$$
$$= 12(1+2.5\times 2+14.1\times 0.8)$$
$$= 24.78\ \text{GPa}$$

$$E_c / E_p = 1 + \left[\frac{15(1-v_p)}{8-10v_p}\right]\left[\frac{\phi_f}{1-\phi_f}\right]$$

$$V_p = 0.68$$

In thermoplastic composites, consolidation and bonding are two important factors to obtaining a good composite. The first step towards this is the *intimate contact* between the reinforcement and the matrix. The second is *autohesion* (*auto*matic + ad*hesion*/co*he-sion*). The following models describe the effects of the processing variables (time, temperature, and pressure) on intimate contact and autohesion.

6.4.1 Intimate Contact

Initially, at the time of placement of reinforcement and matrix are placed together, there are gaps (such as air pockets) between them prior to the application of heat and pressure. Even at high temperatures, the viscosity of thermoplastics is too high to produce the desired degree of flow along the interface. Therefore, thermoplastic matrix composites must actually be deformed to produce intimate contact between adjacent surfaces. Equations used for intimate contact are

$$D_{ic} = b / (w_0 + b_0) \tag{6.5a}$$

$$V_0 = a_0 b_0 = ab \tag{6.5b}$$

where
- D_{ic} = Degree of intimate contact
- a_0 = Initial height of each rectangular element
- a = Spontaneous heights of each rectangular element
- b_0 = Initial width of each rectangular element
- b = Spontaneous width of each rectangular element
- w_0 = Initial distance between two elements
- V_0 = Volume of each rectangular element

$$D_{ic} = \frac{a_0 / a}{1 + w_0 / b_0} \tag{6.6}$$

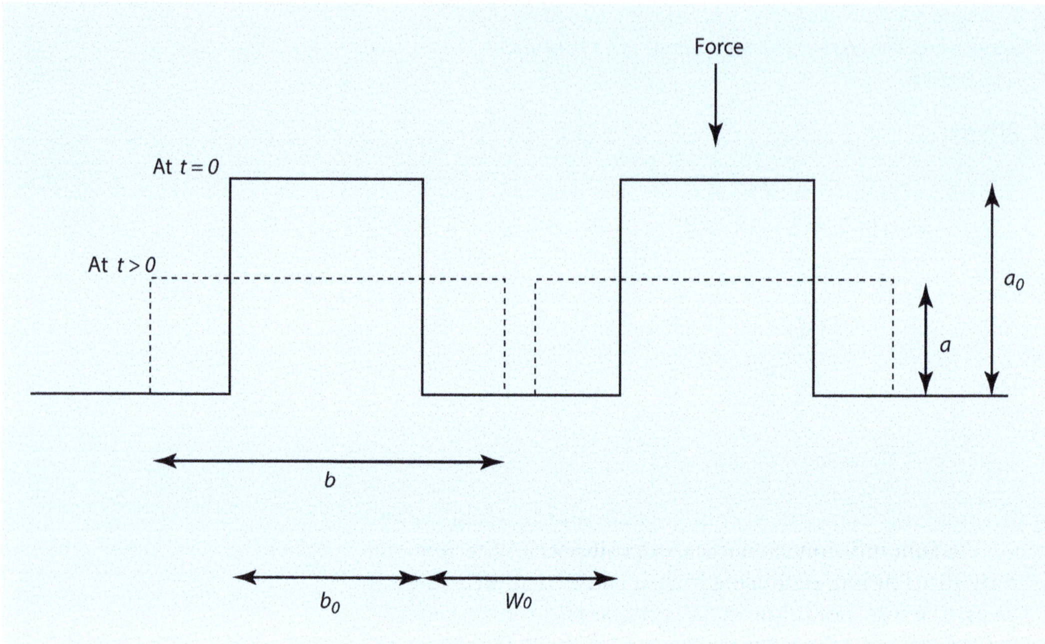

Fig. 6.1 Model for intimate contact

6

The irregular surface in this model is shown by rectangles in Fig. 6.1, and when pressure is applied, these rectangular elements spread along the interface.

Example 6.4 What should the ratio between w_0 and b_0 be to achieve the maximum degree of intimate contact?

Answer

$$D_{\mathrm{ic}} = 1\,(\text{maximum index of intimate contact})$$

$$D_{\mathrm{ic}} = \frac{a_0 / a}{1 + w_0 / b_0}$$

$$a_0 b_0 = a b_0 + a w_0$$

because

$$a_0 b_0 = ab$$

Therefore,

$$ab = ab_0 + aw_0$$

$$\begin{aligned} w_0 / b_0 &= (b / b_0) - 1 \\ &= (b - b_0) / b_0 \end{aligned}$$

In terms of a,

$$w_0 / b_0 = (a_0 - a) / a$$

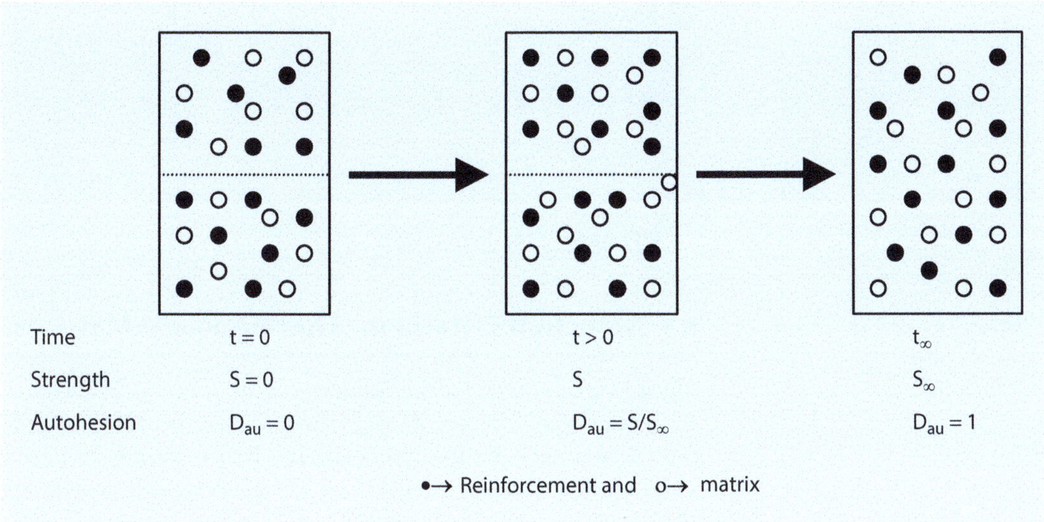

Fig. 6.2 Model for autohesion

6.4.2 Autohesion

Once two interfaces come into intimate contact, a bonding process between the interfaces starts. In the case of a matrix and a particle interface, the bond is formed by *adhesion*, and in the case of two thermoplastic interfaces, the bond is formed by *cohesion*. The combined effect of bonding of all of the interfaces is called autohesion. A model for autohesion is given in Fig. 6.2 where segments of molecules diffuse across the interface as per Eqs. (6.6) and (6.7). The extent of the molecule's diffusion and, hence, the bond strength increases with time.

$$D_{au} = \frac{S}{S_\infty} = 1(\text{Ideal situation}) \quad (6.7)$$

where

- D_{au} = Degree of autohesion
- S = Bond strength at time t
- S_∞ = Ultimate bond strength

Also,

$$D_{au} = k\, t_a^{1/4} \quad (6.8)$$

where

- t_a = Time elapsed in autohesion
- $\kappa = \kappa_0 \exp(-E/RT)$ [Constant that relates to the Arrhenius relation]

Example 6.5 The tensile strength of epoxy resin is 28 MPa, and its bond strength between two aluminium plates is 19.6 MPa. The tensile strength of the aluminium plate is 70 MPa. What is the degree of adhesion and efficiency of this joint?

6

Answer

$$D_{au} = \frac{S}{S_{\infty}}$$
$$= 19.6 / 28$$
$$= 0.7$$

Efficiency of this joint is 70%.

6.5 Selection Criteria for Thermoplastic Matrices

The selection criteria for thermosetting matrices are dependent on the processability factors of resin as well as the processing technology. A summary of those processability factors already has been outlined in ▶ Chap. 1 (▶ Sect. 1.3). The processability factors for thermosetting resins are detailed here.

1. *Properties Driven by Selection Criteria*
   - *Melt Flow Index (MFI)*—MFI is a parameter that indirectly determines the viscosity of a polymer melt. MFI is flow of molten polymer mass measured in grams in 10 minutes under a standard load and temperature. High MFIs favour the percolation of the matrix around fibres and fill the mould cavities faster at lower pressure in both the injection moulding and compression moulding processes [17, 18]. The MFI of various thermoplastic matrices is given in Table 6.2.
   - *Interaction with Reinforcement*—The melt viscosity of a thermoplastic matrix only ensures the percolation and flowability of the matrix around the reinforcement. For example, mineral and glass fibre-reinforced PP did not meet

Table 6.2 MFI of thermoplastics [4]

Serial No.	Thermoplastic	MFI (g/ 10 min)
1.	Polypropylene (PP)	5–60
2.	Polyethylene (PE)	2–70
3.	Acrylonitrile butadiene styrene (ABS)	5–25
4.	Nylon 6	5–80
5.	Nylon 6,6	5–80
6.	Polyacetal	3–70
7.	Polycarbonate (PC)	2–30
8.	Polyethersulfone	7–18
9.	Polyether ether ketone (PEEK)	1–20
10.	Polypropylene oxide	7–30
11.	Polybutylene terephthalate (PBT)	20–100
12.	Polymethyl methacrylate (PMMA)	2–40

Fig. 6.3 Mineral and glass fibre-reinforced PP tub for a washing machine [19]

the specifications for washing machine tubs due to a lack of flowability. This was changed by adjusting of the grade of PP for better flowability, good interphase properties, and superior hydrolysis resistance. This innovation revitalized the washing machine tub design, as shown in Fig. 6.3.

- *Chemical Interaction with Reinforcement*—Chemical interaction between a thermoplastic resin and the reinforcement are usually not as strong as in the case of thermosetting matrices. However, coupling agents on the reinforcement play a major role in this property. Thermoplastic resin must be selected based on its reactivity with the surface of the reinforcement to make a good interphase.
- *Molecular Weight and Its Distribution*—Thermoplastic matrices have high molecular weights, and the molecular weight distribution (MWD) in most polymers with a high molecular weight is wide. The molecular weight in thermoplastics, such as polyolefins, ranges from 3.1×10^6 to 5.7×10^6 g/mol. In highly crystalline polymers, the MWD is narrow, and that is why they have sharp melting points. The molecular weight distribution is quite wide in amorphous and semi-crystalline matrices, and it ranges from 1.5 to 4.0. High MWD causes the initial softening of a thermoplastic followed by its melting. This occurs due to the melting of low molecular weight moieties in the polymers and the delayed softening of high molecular weight moieties.
- *Thermal Stability*—The thermal stability of thermoplastic resins is limited. Only high-temperature thermoplastics (polyether ether ketone (PEEK), polyetherimide (PEI), polyphenylene sulphide (PPS), polyamide-imide (PAI),

6

etc.) are resistant to high temperatures. In thermoplastic matrices, T_g is different from the melting and processing temperatures, as explained in ▶ Chap. 1. T_g is much lower than the melting temperature (T_m), and the melting temperature is lower than the processing temperature (T_p), i.e. $T_g < T_m < T_p$.

- *Hygrothermal Stability*—The hygrothermal stability of thermoplastic resins is good. Ester-based thermoplastics and amides (such as nylon) are susceptible to moisture absorption. Polyolefins and high-temperature polymers (PEEK, PEI, PPS, etc.) offer good resistance to moisture.
- *Post-Processing Operation*—Post-processing operations are minimal in thermoplastic composites. Embossing and engraving are easily done on the surface of thermoplastic matrices due to their melting characteristic. The reactivity of the surface of thermoplastic composites makes it compatible to printing too.
- *UV Resistance*—Exposure of thermoplastic composite products has been a major criteria for the selection of a thermoplastic matrix. UV radiation from 290 to 400 nm is detrimental to thermoplastic matrices—specifically polyolefins. Ester-based thermoplastics (polymethyl methacrylate (PMMA), polycarbonates, etc.) are inherently resistant to UV radiations. The mechanism for UV resistance for ester-based thermoplastics is the absorbance of UV radiation and the dissipation of UV energy as heat energy through the oscillation of a double bond between –C = O and –O, as explained in ▶ Chap. 5.
- *Fire Resistance*—Fire resistance has been a criterion for the selection of a thermoplastic matrix. Polyvinyl chloride (PVC) is inherently fire-resistant due to the presence of a chloride group. Apart from PVC, all thermoplastics containing an ether group (–O–) are fire-resistant. In cases where fire resistance is a must, bismuth oxides or antimony oxides are added to improve fire resistance.

Point to Ponder...

***P**olycarbonates are used in headlamp covers for automobiles due to their transparency, high-impact strength, and UV resistance. Polyester films are used for making signages and displays for outdoor advertisements due to their UV resistance.*

2. *Processing Technology Driven by a Selection Criteria*
   - *Compatibility with the Form of Resin*—Thermoplastic resins are available in three forms: granules, films, and sheets. They are always in solid form; therefore, their processing requires external pressure—apart from heat—for compaction of the composite. *Granules* are used to make injection-moulded composite parts. Granules are available with different percentages of reinforcements of glass fibres, carbon fibres, etc. *Films* and *sheets* are primarily used for the fabrication of composites using a compression moulding process.
   - *Flow of a Matrix*—The flow of a matrix is important for the percolation of the matrix and good flow is required for the injection moulding process. Flow requirements are dependent on the type of processing technology used and are also related to the mould design.

- *Compatibility of Mould Design with a Matrix*—Mould design dictates the path to be travelled by a polymer melt, the volume it needs to fill, what is the filling time, etc. A suitable polymeric matrix needs to be selected to fill the mould cavity optimally in an injection moulding process and to ensure percolation of the resin in a compression moulding process.
- *Production Rate*—Production rates have always been one of the guiding rules for choosing a processing method. In thermoplastic composites, very high production rates are achieved for short fibre-reinforced thermoplastic composites using an injection moulding process. The rate of production may be up to 20 products per minute, depending upon the volume of the product, capacity of the injection moulding machine, mould design, material used, processing parameters, etc. In compression moulding processes, very high production rates are not achievable. However, with multi-daylight compression moulding machines, high production rates are feasible, enhancing the production rate. The major difference between the production rate of an injection moulding process and a compression moulding process is the process cycle time. The process cycle time is 30–60 seconds on an average in an injection moulding process, whereas in a compression moulding process, this time may range from 2 to 6 hours.
- *Cost of Production*—The cost of production is usually low in thermoplastic composites due to high production rates. Also, the processes used in thermoplastic composites are more automated; therefore, chances of rejection of the products moulded are very less. The rejects do have scrap value, and this will offset the production cost.
- *Product Size and Shape*—Product size and shape has a bearing in the selection of the matrix because it has to travel a path to fill the mould cavity in an injection moulding process and the thermoplastic resin needs to percolate into the entire mould in a compression moulding process. For large products, crystalline or semi-crystalline polymeric matrices (such as nylons, polypropylene, and polyethylene) are preferred. Amorphous polymeric matrices (such as PC and PMMA) are also used for small- to medium-sized composite products in an injection moulding process. High-temperature semi-crystalline matrices (such as PEEK and PEK) need thermal management in order to control the heat losses; therefore, they are not preferred for large composite products using an injection moulding process. Similarly, high-temperature amorphous matrices are also used for small composite products due to thermal management considerations and low flowability. Compression moulding processes may be used for all sizes of products having any kind of matrix.

6.6 Processability of Commonly Used Thermoplastic Matrices

Thermoplastic resins first melt and then solidify. Some of the commonly used thermoplastic resins and their processability factors are described here.

Point to Ponder…
There are variety of grades in nylon, such as Nylon 6; Nylon 6,6; Nylon 6,12; Nylon 6,11; etc. *However, the most commonly used grades are Nylon 6 and Nylon 6,6.*

6

1. *Nylon*—Nylon 6 and Nylon 6,6 are the major engineering plastics being used in short fibre-reinforced thermoplastic composites. Nylon 6 is manufactured from a ring-opening reaction of caprolactam, and Nylon 6,6 is manufactured from a polycondensation reaction of adipic acid with hexamethylenediamine. Water from the reaction comes as a condensate. The properties of Nylon 6,6 are superior to Nylon 6. Nylon is made available in the form of glass-reinforced granules, homopolymer films, and homopolymer sheets. The general properties of Nylon 6 are given in Table 6.3.

The major processability factors of nylons are listed here.

- The melt viscosities are very low (i.e. 500 cps) due to its high crystalline nature.
- The processing temperature of nylons is quite high (250 °C). Care must be taken to thermally process less stable natural fibre-reinforced composites.
- They are suitable for prepregging, and carbon, glass, and Kevlar fibre-based nylon prepregs are used for vacuum pressure moulding.
- Nylon may be processed using either injection moulding or compression moulding processes.
- They have a low shrinkage (<0.5%).
- Nylons are hygroscopic in nature due to the presence of an amide group.

Point to Ponder…
Polypropylene is one of the wonder materials in the manufacture of orthoses for physically challenged persons. PP's low density, good mechanical properties, lack of reaction from the human body, and excellent abilities to be both moulded and machined make it the best material for this purpose.

2. *Polypropylene*—Polypropylene (PP) is a commodity plastic that is made into engineering plastic after reinforcement. It is manufactured by addition reaction of propylene gas using a Ziegler–Natta catalyst. The bulk manufacturing process for PP is emulsion polymerization. PP is available in three basic forms: isotactic (highly crystalline in nature), syndiotactic

Table 6.3 Properties of Nylon 6,6

Serial No.	Property	Value
1.	Density (g/cm^3)	1.2
2.	Tensile strength (MPa)	65
3.	Young's modulus (GPa)	4
4.	Elongation at break (%)	80
5.	Impact strength, notched (KJ/m)	0.07
6.	Glass transition temperature (°C)	45
7.	Processing temperature (°C)	240–350

(semi-crystalline in nature), and atactic (amorphous in nature). Apart from these three forms, the copolymer of PP used with ethylene is also popular and is also amorphous in nature. Isotactic PP is used in composite industries. PP is available in the forms of granules, films, sheets, and fibres. The general properties of PP are given in Table 6.4.

The major processability factors for PP resins are the following:

- PP is highly crystalline in nature; therefore, it has a melt viscosity that is low and is easily mouldable.
- It is used in thermoplastic matrix using both injection moulding and compression moulding processes.
- PP goes under thermal degradation fast at a high temperature; therefore, the processing temperature needs to be controlled.
- For UV stabilization, hindered light amine stabilizers (HALS) are added to polypropylene in small quantities.
- It is chemically inert. Therefore polypropylene is suitable for many applications.
- Polypropylene has a very low shrinkage (approximately 0.2%).

3. *Polycarbonate*—Polycarbonate (PC) is a high-impact strength thermoplastic that is used for applications, such as bullet-resistant glasses. PC is manufactured using condensation reaction of bisphenol 'A' and carbonyl chloride. The water of reaction comes out as a condensate. It is available in the form of granules, sheets, and films. The fibres of amorphous polymers such as PC cannot be drawn. The general properties of polycarbonates are shown in Table 6.5.

Point to Ponder...
The morphology of PC has interstices (air gaps). Therefore, PC allows the tightening of screws in it, like wood.

The major processability factors for polycarbonate resins are the following:

- PC is an amorphous thermoplastic having very good transparency.
- Polycarbonates are used for optical applications.
- Polycarbonates are used for high-impact resistant applications.
- PC contains an ester group; therefore, it is resistant to UV radiation.

Table 6.4 Properties of polypropylene

Serial No.	Property	Value
1.	Density (g/cm^3)	0.94
2.	Tensile strength (MPa)	33
3.	Young's modulus (GPa)	1.4
4.	Elongation at break (%)	150
5.	Impact strength, notched (KJ/m)	0.07
6.	Glass transition temperature (°C)	0 (Isotactic)
7.	Processing temperature (°C)	210–290

Table 6.5 Properties of polycarbonate

Serial No.	Property	Value
1.	Density (g/cm^3)	1.2
2.	Tensile strength (MPa)	65
3.	Young's modulus (GPa)	2.4
4.	Elongation at break (%)	110
5.	Impact strength, notched (KJ/m^2)	27
6.	Glass transition temperature (°C)	147
7.	Processing temperature (°C)	240–310
8.	Refractive index	1.585

Table 6.6 Properties of ABS

Serial No.	Property	Value
1.	Density (g/cm^3)	1.04
2.	Tensile strength (MPa)	22
3.	Young's modulus (GPa)	1.6
4.	Elongation at break (%)	57
5.	Impact strength, notched (J/m)	106
6.	Glass transition temperature (°C)	104
7.	Processing temperature (°C)	180–200

- PC is hygroscopic in nature due to the presence of an ester group. PC granules and sheets need drying to remove moisture before moulding.
- Polycarbonates are both injection and compression mouldable for the fabrication of thermoplastic composites.
- PC has a low shrinkage (>0.5%).

4. *Acrylonitrile Butadiene Styrene (ABS)*—ABS is a terpolymer and is amorphous in nature. It is one of the most important engineering thermoplastics, and it is often used as matrix with short glass fibres as the reinforcement. It is available in the form of granules and sheets. The general properties of ABS are given in Table 6.6.

The major processability factors for phenolic resins are listed here:

- ABS is an amorphous thermoplastic. As a matter of fact, all copolymers, terpolymers, etc. are amorphous in nature.
- It is often used as matrix in an injection moulding process. Although ABS also can be used in a compression moulding process with fabric reinforcement.
- Acrylonitrile butadiene styrene has a low shrinkage (<0.5%).

Table 6.7 Properties of PBT

Serial No.	Property	Value
1.	Density (g/cm^3)	1.4
2.	Tensile strength (MPa)	60
3.	Young's modulus (GPa)	2.4
4.	Elongation at break (%)	50
5.	Impact strength, notched (KJ/m^2)	27
6.	Glass transition temperature (°C)	40
7.	Processing temperature (°C)	230–250

5. *Polybutylene Terephthalate (PBT)*—PBT is one of the rarest crystalline polyester-based thermoplastic. It is available in the form of granules, fibres, sheets, and films. It is manufactured using a polycondensation reaction of purified terephthalic acid (PTA) or dimethyl terephthalate (DMT) with 1,4-butanediol (BDO). Water from the reaction comes out as condensate. The general properties of PBT are shown in Table 6.7. The major processability factors for PBT are the following:
   - PBT has good flowability due to its low melt viscosity.
   - It provides a good surface finish to the moulded compound.
   - It is often used as matrix with short glass fibres in an injection moulding process.
   - PBT is used as matrix with alumina whiskers reinforcing thermoplastic composites.
   - PBT makes good blends with many other thermoplastics.
   - Polybutylene terephthalate has a low mould shrinkage (> 0.7%).
6. *Polyether ether ketone (PEEK)*—PEEK is a high-performance engineering thermoplastic. It is made from a reaction of 4,4′-difluorobenzophenone or 1,4-bis(4-fluorobenzoyl) benzene with hydroquinone in the presence of an alkali carbonate (such as anhydrous potassium carbonate) and in a polar aprotic solvent (such as diphenyl sulphone (DPS)) at a temperature close to the melting point of the polymer. The general properties of PEEK are given in Table 6.8.

The major processability factors for PEEK are the following:
   - PEEK has enormous potential for applications in harsh environments due to its chemical resistance. It is inert to all chemicals except concentrated sulphuric acid.
   - It is used as a high-temperature thermoplastic matrix.
   - Due its semi-crystalline nature, PEEK may be moulded as easily by injection moulding as by compression moulding processes.
   - Due to its availability in the form of fibres, fabrics, films, and sheets, PEEK is compression moulded with a good volume fraction of reinforcement.
   - PEEK has a low shrinkage of 0.7%.

Table 6.8 Properties of PEEK

Serial No.	Property	Value
1.	Density (g/cm^3)	1.32
2.	Tensile strength (MPa)	97
3.	Young's modulus (MPa)	3.6
4.	Elongation at break (%)	60
5.	Impact strength, notched (J/m)	83
6.	Glass transition temperature (°C)	143
7.	Processing temperature (°C)	350–400

Table 6.9 Properties of PEI

Serial No.	Property	Value
1.	Density (g/cm^3)	1.27
2.	Tensile strength (MPa)	100
3.	Young's modulus (MPa)	2.7
4.	Elongation at break (%)	110
5.	Impact strength, notched (J/cm)	0.27
6.	Glass transition temperature (°C)	216
7.	Processing temperature (°C)	290–350

7. *Polyetherimide (PEI)*—PEI is high-performance engineering thermoplastic. It is amorphous in nature. Polyetherimide is manufactured through complex and multistep reactions of bisphenol 'A', phthalic anhydride, and aromatic amine. The general properties of PEI are shown in Table 6.9.
   - The major processability factors for PEI resins are listed here:
   - The MFI is low; therefore, the flowability of PEI is restricted. It requires high pressures for injection moulding or percolation through fibres.
   - PEI is compatible for both injection and compression moulding processes.
   - It is used as a high-temperature thermoplastic matrix.
   - The films of PEI are made using a solvent-based (i.e. dimethylacetamide) casting process.
   - PEI has a low shrinkage (>0.5%).

6.7 Selection Criteria for Reinforcement

Reinforcements, as we know, are the main building blocks of a composite because they are the load-bearing members. Reinforcements are responsible for most of the major properties of composite. Their properties, size, and shape are important for processability. A table showing the processability of reinforcement and matrices vis-á-vis to the processing methods was presented in Table 5.2 of ► Chap. 5. The selection criteria for the reinforcement of thermosetting composites are dependent on the processability factors of reinforcement as well as the processing technology. The processability factors for reinforcement are detailed next.

6.7.1 Properties Driven Selection Criteria

- *Form of Reinforcement*—In thermoplastic composites, all forms and shapes of reinforcements, as described in earlier sections, are used. Usage of reinforcement depends upon the moulding process described in Table 6.10.
- *Wettability of Reinforcement*—The surface of reinforcement must have a high surface energy in order to allow the resin to wet it. The wettability of the reinforcement ensures a good interphase between the reinforcement and the matrix. This will result in a good composite.
- *Chemical Interaction with Resin*—The hydrophobic surface of reinforcement will allow the formation of a chemical bond with an organic matrix. This will yield into a good interphase. The reinforcement must be selected based on its capability of having a good chemical interaction with matrix.
- *Fire Resistance*—This is an important parameter from the point of view of safety. Mostly synthetic reinforcing fibres are fire safe—except for UHMWPE, PP, and PEEK. Natural fibres are susceptible to fire hazards, so care must be taken that the processing temperature must be less than their ignition temperature.

Table 6.10 Processability of reinforcement and thermoplastic matrices vis-á-vis their processing methods

Serial No.	Reinforcement	Processing method	Matrices
1.	Synthetic fibres (glass/carbon/aramid)		All thermoplastic matrices
	(a) Short fibres/staples/whiskers	Injection moulding	
	(b) Long fibres	Compression moulding	
2.	Natural fibres		All thermoplastic matrices (except high-temperature matrices, such as PEEK, PEI, PES, PEK, or PAI)
	(a) Short fibres/staples	Injection moulding	
	(b) Long fibres	Compression moulding	

- *Thermal and Hygrothermal Stability*—The thermal stability of synthetic fibres (such as glass, carbon, and Kevlar) is good, and they can be used under any thermal processing. Natural fibres are thermally less stable, and care must be taken for their processing, as mentioned previously. Moisture affects all fibres, and the ingress of moisture in Kevlar deteriorates its mechanical properties. In glass fibres, moisture causes depletion of the coupling agent. In carbon fibres, moisture affects the polymeric sizing of the polyimide or the epoxy over it. Natural fibres absorb moisture; thus, their mechanical properties come down.
- *Post-Processing Operations*—The role of reinforcement and its machinability is important for post-fabrication operations. Kevlar reinforced composites are difficult to machine, and they require water jet or laser cutting operations for smooth cutting edges. Kevlar composites having a fibril structure will cause deformation from fibrillization of the fibres and give a rough surface on mechanical machining by standard cutting tools (hacksaw, milling cutter, turning tools, etc.) The processability of composites does not end with striking the correct combination of raw materials and processing methods to just get the product moulded. Both the designer and the process engineer must be prudent enough to plan in the processing of the composite products for post-fabrication finishing. Thus, the raw materials, moulding surfaces, and processing technologies need to be chosen in such a manner that they ensure a minimum of post-fabrication operations for moulded composite products.
- *UV Resistance*—UV radiations do not affect natural fibres and most of the synthetic reinforcing fibres—except polyolefin fibres, such as PP and UHMWPE.

6.7.2 Processing Technology Driven by Selection Criteria

Primarily, there are only three processes (i.e. injection, compression, and pressure bag moulding), which are used for fabrication of thermoplastic composites. Of late, a thermoplastic prepreg-based filament winding process also has been developed.

- *Compatibility with the Form of Reinforcement*—Short fibres, staples, whiskers, and particles only can be used as the reinforcement in an injection moulding process. For example, the Motorola Ultra smartphone has a rear housing made of a Kevlar short fibre-reinforced PC/PET blend, as shown in ◘ Fig. 6.4. In a compression moulding process, reinforcement in any form may be used. Roving and tow preregs are used in filament winding technology, but thermoplastic filament winding is tedious because it involves many processing steps and stringent control of processing parameters. The pressure bag method also employs fabric prepregs.

Fig. 6.4 Rear housing of Motorola made out of Kevlar reinforced PC/PET composite [20]

- *Flow of a Matrix*—Short fibres/whiskers/particle reinforcements increase the viscosity of the mixture of matrix and reinforcement. The flow of a matrix is essential for the resin to percolate throughout the reinforcement so that complete wettability of the reinforcement by resin is achieved.
- *Production Rate*—Injection moulding supports high production rates. Other production methods are batch-forming processes; therefore, the production rates are comparatively lower.
- *Cost of Production*—The cost of production is directly linked with the rate of production and cost of raw materials. The selection of the right kind of reinforcement for a required service and the service life of the composite is the key.

6.8 Methods to Improve Processability in Thermoplastic Composites

The methods and mechanisms to improve the processability in thermoplastic composites are adsorption and wetting, chemical bonding, mechanical adhesion, interdiffusion, and electrostatic attraction. These have been described in detail in ▸ Chap. 5. Apart from these methods and mechanisms to improve the processability, crystallization of the matrix on the fibre is also carried out. Moderate crystallization creates an inherently strong interphase region because of the intimacy of the crystal growth pattern on the fibre surface. This is shown schematically in Fig. 6.5. Therefore, crystallization can either decrease or increase adhesion depending on the relative effects of interdiffusion and crystalline intimacy.

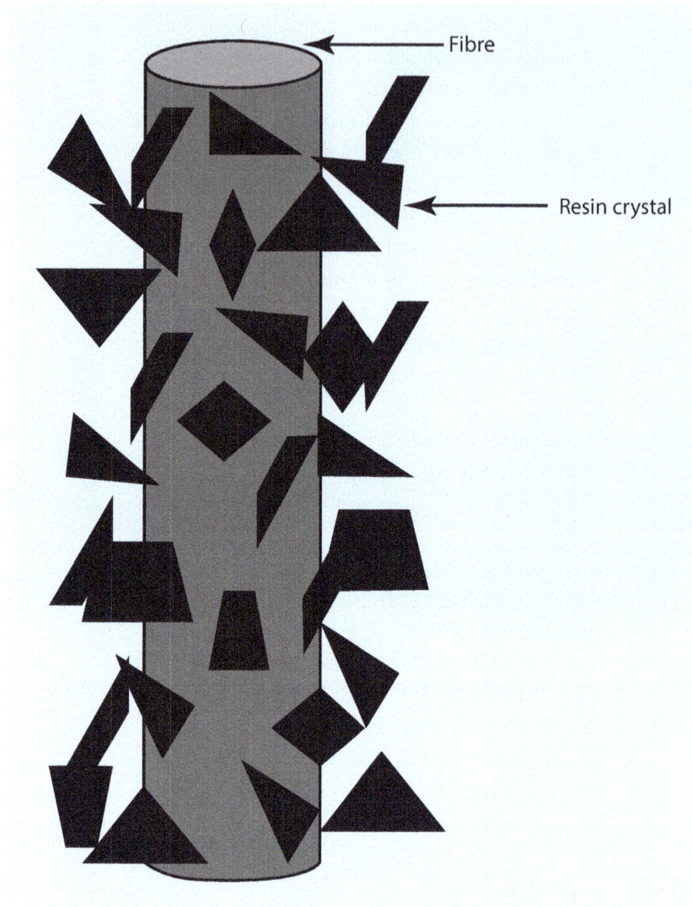

Fig. 6.5 Schematic presentation of the crystallization of a matrix on the fibre surface

6.9 Processability of Thermoplastic Green Composites

Thermoplastic composites offer more environmentally friendly 'green composites' due to one or more added advantage over thermosetting matrix composites in the form of their recyclability. *Green composites* do not cause any harm to the environment during their processing and their disposal after their useful life is over. In this classification of composites, major composites are those that use natural fibres as reinforcement and/or have biodegradable matrices. Green composites have created interest worldwide due to their environment friendliness. These composites are not new. In fact, their knowledge has existed with mankind from time immemorial. In ancient times, straw- or bamboo-reinforced mud houses were built (as mentioned in ▸ Chap. 1), and natural fibres (such as coir or jute) were used as reinforcement in the construction of buildings. The densities of these fibres are in the range of 1.15–1.55 g/cm^3, which supports their processability using common processes and having matrices that are processable within the limit of thermal degradation of natural fibres. Rubberized cotton, which was earlier used in bicycle tyres, is an apt example of a natural fibre-reinforced polymeric 'green' composite.

Example 6.6 Discuss specific applications of thermoplastic green composites.

Answer

Bamboo or coir fibre-reinforced polyethylenes and polypropylenes are being used as door panels in civil construction industries. Bamboo fibre-reinforced PLA is being used for various applications for complete biodegradability.

However, the use of biodegradable fillers (such as starch or wood flour) in conventional thermoplastic or thermosetting matrices increases the scope of biodegradable matrices. It is to be noted that the bulk density of natural fibres is quite high, and this makes their weight percentage in the composite quite low.

True green composites are manufactured by reinforcing natural fibres in biodegradable polymeric matrices, such as PLA, PCL, or PGA. These fibres are thermoplastic in nature and can be processed using various methods, as applicable to other thermoplastics. The processability of biodegradable composites is carried out in a closed mould processing. The processability parameters for biodegradable composites are as follows.

6.9.1 Processability of Natural Fibre-Reinforced Thermoplastic Matrix Composite

The processability of these composites is governed by the following factors:

- *Interphase in the Composite*—The hydrophilic nature of natural fibres restricts their wetting by hydrophobic polymeric matrices, which has been discussed in ► Chap. 5. However, reactive thermoplastic matrices (Nylon, PMMA, ABS, etc.) make a reactive bond with natural fibres, yielding a good composite.
- *Hygroscopic Nature of Natural Fibres*—Natural fibres require adequate drying before processing with thermoplastic matrices. The moulding of thermoplastic composites is carried out in closed moulds. (This has been discussed in more detail in ► Chap. 7.) Therefore, there are remote possibilities for moisture or entrapped gases to escape. The ingress of moisture in the composite reduces its properties as well as its aesthetic appearance.
- *Low Strength*—The low strength of natural fibres restricts their processability using filament winding, tape lay-up, or pultrusion methods.
- *Use of High-Temperature Matrices*—Due to the low thermal stability of natural fibres, their processability as reinforcement in high-temperature matrices (PEI, PEEK, polyimides, etc.) is not feasible.

- *Service Environment*—A service environment having high temperatures or high moisture is not suitable for natural fibre-reinforced polymeric composites.
- *Safety*—The processability of these composites does not cause any health hazard due to the nontoxicity of the natural fibres.

6.9.2 Processability of Natural Fibre-Reinforced Biodegradable Matrix Composites

Natural fibre-reinforced biodegradable matrix composites are the true green polymeric composite. Biodegradable polymers are used as matrices with natural fibres as reinforcements. In order to make a synthetic matrix biodegradable, biodegradable materials (such as starch or wood flour) are also mixed in the polymeric matrix. The processability of these composites is governed by the following factors:

- *Interphase in the Composite*—The hydrophilic nature of natural fibres restricts their wetting by hydrophobic polymeric matrices, as mentioned in ► Sect. 4.5.1.
- *Brittleness of Biodegradable Polymers*—PLA is one the most important and common biodegradable polymers. It is quite brittle; therefore, processability of its composite is difficult. Often, a need is felt for plasticizing it before processing. The problem of brittleness is also compounded by the low-impact strength of natural fibres.
- *High-Temperature Processing*—High-temperature processing of a biodegradable matrix (PLA) restricts the processability of natural fibres with low thermally stability.
- *High Cost of Biodegradable Polymers*—The cost of biodegradable polymers is quite high in comparison to their thermoplastic counterparts. This restricts their use in many common applications.
- *Durability*—The durability of natural fibre-reinforced biodegradable composites is limited by the durability of the biodegradable polymer in the operating environment.
- *Service Environment*—Service environments having high temperatures or high moisture content are not suitable for natural fibre-reinforced polymeric composites, as mentioned in ► Sect. 5.5.
- *Safety*—The processability of these composites does not cause any health hazard due to the fact that natural fibres are nontoxic, as mentioned in ► Sect. 5.5.

6.9.3 Processability of Synthetic Fibre-Reinforced Biodegradable Matrix Composites

Synthetic fibres (such as glass fibre-reinforced PLA composites) are common, and their commercial development will be visible in the near future. These composites offer the best processability in terms of the compatibility of raw materials, ability to form interphase, and adaptability to different processing technologies. Composites from biodegradable matrices do not pose any problem from degradation due to moisture or high temperature. However, some of the parameters that govern their processability are given here:

- *Brittleness of Biodegradable Polymers*—PLA is quite brittle; therefore, the processability of its thin section is difficult. Thin composite sections tend to break at low shocks that are encountered in an injection moulding process.
- *High Cost of Biodegradable Polymers*—The cost of processing biodegradable polymers is quite high in comparison to their thermoplastic counterparts. This restricts their use in many common applications, as mentioned in ▶ Sect. 5.5.
- *Durability*—The durability of natural fibre-reinforced biodegradable composites is limited by the durability of the biodegradable polymer in the operating environment, as mentioned in ▶ Sect. 5.5.

To Be Remembered

1. Thermoplastic matrices are always solid to start with.
2. Thermoplastic matrices are not processed at room temperature; they are only processed at high temperatures.
3. The melt viscosities of thermoplastic matrices are higher when compared to thermosetting matrices.
4. Spectra fibres are not chemically treated and are transformed into composites with low-temperature matrices.
5. Moulding of thermoplastic composites does not involve any curing methods. Rather, it involves the melting of the matrix and its solidification as a function of heat.
6. Intimate contact takes place between the reinforcement and the matrix, whereas autohesion is coalescence within the matrix in thermoplastic composites.
7. Melt viscosity of thermoplastic matrices only ensures the percolation of the matrix around the reinforcement.
8. High MFIs indicate a low viscosity of the matrix.
9. Short fibres, particles, and whiskers are used as reinforcement in an injection moulding process.
10. All of the forms of reinforcement (fibres, fabrics, particles, whiskers, staples, etc.) could be used in compression moulding processes.
11. The major processes for making thermoplastic composites are injection, compression, and pressure bag moulding.
12. Thermoplastic prepreg rovings and tows are being used for filament winding process.
13. Nylon has sharp melting point due to its high crystallinity.
14. PEEK and PEI are mostly used for high-temperature thermoplastic matrices.
15. PEEK is a chemically resistant thermoplastic—except when used with concentrated sulfuric acid.
16. PP is the least dense thermoplastic matrix used in thermoplastic composites.

6

Exercises

1. Calculate the degree of autohesion in a glass-reinforced polypropylene composite having an experimental value of ILSS of 9 MPa. The ideal ILSS of this composite is 12 MPa.
2. The initial height and width of an undulation of fabric are 0.1 mm and 0.25 mm, respectively. The initial distance between two undulations is 1 mm. After moulding the composite using the same fabric, the final height of undulation is observed to be 0.05 mm. Calculate the degree of intimate contact.
3. Compare the modulus of a talc-filled polypropylene composite with that of a glass fibre-reinforced polypropylene. The volume fractions of talc and glass fibre are 0.4 and 0.5, respectively. The moduli of polypropylene and glass fibre are 3 GPa and 70 GPa, respectively.
4. Calculate Poisson's ratio for filler-reinforced composites. The volume fraction of filler is 0.25, and Young's modulus is 70 GPa.
5. Why is a reinforcement with low thermal stability not preferred in high-temperature thermoplastic composites?
6. Outline the challenges encountered in making natural fibre-reinforced thermoplastic composites.
7. Why are 'green composites' generally thermoplastic but not thermosetting in nature?
8. Which is the most common chemical group that makes a plastic inherently fire resistant?
9. Calculate the load-bearing area of a cylindrical particulate composite. The length of the cylindrical portion of the particulate filler is 3 mm and the Φ_f value is 0.2. The dimension of a particle may be taken in millimetres for the sake of calculation.
10. What are the post-processing operations in thermoplastic composites? Outline the significance of post-processing operations also.

References

1. Shaffer, G.D., "An Archaeomagnetic Study of a Wattle and Daub Building Collapse," *Journal of Field Archaeology*, 20 (1) (1993), p. 59.
2. Pukanszky, B., Tudos, F., Kolarik, J., and Lednicky F., "Ternary Composites of Polypropylene, Elastomer, and Filler: Analysis of Phase Structure Formation", *Poly. Compos.*, 11 (1990), p. 98.
3. Brydson J.A., *Plastics Materials*, Sixth Edition, Butterworth Heinemann, USA, 1995.
4. *Plastic Molding Tutorials: #107 Melt Flow Rate (MFR) of Plastics*, http://www.misumitechcentral.com/tt/en/mold/2012/01/107.html
5. Guth, E. and Smallwood, H. J., "Limiting Law of the Reinforcement of the Rubber", *Appl. Phy.*, 15 (1944), p. 758.
6. Hopgarten, H., *Fibre Science and Technology*, 11 (1978), p. 67.
7. Dwivedi, M., Alam, S., and Verma, G.L., "Effect of Nano-barium Titanate on Mechanical, Electrical, and Solubility Performances of Ferrite-Filled Poly (ether ether ketone) (PEEK) Composites," *Polymer International*, 54 (2) (2004), p. 401.
8. Varma, I.K., Gupta, A., Sangita, and Varma, D.S., "Addition Polyimides. I", *J. Poly Sci.*, 28 (1983), p. 191.
9. Varma, I.K., Sangita, and Varma, D.S., "Addition Polyimides. IV. Effect of Structure on Thermal Characteristics", *J. Poly Sci.*, 29 (1984), p. 2807.

10. Siegmann, A. and Turi, E.A., *J. Macromol. Sci.- Phys.*, B10(4) (1974), p. 689.
11. Corten, H.T., *Engineering Design for Plastics*, Baer E., Ed., Reinhold Publishing Corporation, New York, 1964.
12. Russo, P., Cimino, F., Acierno, D., Giovanni, L., and Petrarca, C., "Poly(Butylene Terephthalate) Based Composites Containing Alumina Whiskers: Influence of Filler Functionalization on Dielectric Properties," *International Journal of Polymer Science*, 2014 (2014), p. 1.
13. Dwivedi, M., Gaur, K. K., Alam, S., Bhatnagar, N., and Ghosh, A. K., "Processing of Polyetherimide/Cloisite 30B Nanocomposite by Vibration Casting Method," *International Polymer Processing*, XXIX, 2014, p. 635.
14. Dwivedi, M., Dixit, A., Alam, S., and Ghosh, A. K., "Dielectric and Tensile Behaviour of Nanoclay Reinforced Polyetherimide Nanocomposites," *Journal of Applied Polymer Science*, 122 (2) (2011), p. 1040.
15. Young, R., *Handbook of Fibre Glass and Advanced Plastics Composites*, Lubin, G., ed., Van Nostrand Reinhold Company, New York, 1969.
16. Gandhi, K.S. and Burns, R., *"Rheological Propereties of Glass-reinforced Dough Moulding Compounds", Trans. Rheo Soc.*, 20 (1976), 489.
17. Suddel, B.C. and Evans, W.J., "Natural fibre composites in automotive applications," *Natural Fibres, Biopolymers, and Biocomposites*, (Eds. Mohanty, A.K., Misra, M., and Drzal, L.T.), CRC Press, Taylor & Francis, 2005.
18. Sabir, M.I., Xu, X. and Li, L., "A review of biodegradable polymeric materials for bone tissue engineering applications," *J. Mater. Sci.*, 44(21) (2009), p. 5713.
19. "Reiforced thermoplastic composite for washing machnine tub," *FRP Today*, 14 (3), March 2014, p. 18.
20. "All Thermoplastic Composites Smartphone Back Covers," *FRP Today*, 14 (3), March 2014, p. 22.

Processability in Open Mould Processing of Polymeric Composites

A. K. Ghosh, M. Dwivedi, *Processability of Polymeric Composites*,
https://doi.org/10.1007/978-81-322-3933-8_7

7

Polymeric composites offer the largest processing window among all of the engineering materials due to their wide range of available raw materials and vast horizons of processing technologies for the production of specific composites. When put together, this provides great flexibility in their processing. Advanced composites are also amenable to post-fabrication processing, such as machining and painting. The processing methods and parameters of polymeric composites depend upon the type of raw materials to be processed. The most common processing methods used for making thermoplastic (TP) matrix composites and thermosetting (TS) matrix composites are provided in ◘ Table 7.1.

Processing composite is a vast subject area and has immense academic and commercial interest all over the world. The selection of a manufacturing process depends on a number of factors, such as end use of the product, raw materials, size and geometry of the product, production volume, and cost of processing. On one hand, there are a few common processing technologies (such as compression moulding, injection moulding, extrusion, prepregging, autoclaving, pressure bagging, and vacuum forming) that are common to both thermosetting and thermoplastic resins. On the other hand, there are certain processing technologies that are specific to either thermoplastics or thermosets.

◘ **Table 7.1** Processing methods for thermoplastic (TP) and thermosetting (TS) matrices composites

Serial no.	Processing method	Type of matrix	Type of reinforcement
1.	Injection moulding	TP	Particulate, short fibre
2.	Thermoforming	TP	Particulate
3.	Extrusion	TP	Particulate, short fibre
4.	Blow moulding	TP	Particulate
5.	Calendaring	TP	Particulate
6.	Compression moulding	TP/TS	Particulate, short fibre, long fibre
7.	Autoclaving	TS	Short fibre, long fibre
8.	Contact layup	TS	Short fibre, long fibre
9.	Matched die moulding	TS	Short fibre, long fibre
10.	Filament winding	TS	Long fibre
11.	Reaction inj. moulding	TS	Short fibre
12.	Resin transfer moulding	TS	Long fibre

The fabrication of thermoplastic composites needs special attention due to their high viscosities, high surface tension, inertness, etc., whereas thermosetting resins have low surface tension and good reactivity. Composite manufacturing processes may be divided into two types: open moulding and closed moulding. The processes in *open moulding* are contact layup, spray layup, filament winding, automated tape winding, and casting. With open moulding, the laminate is exposed to the atmosphere during the fabrication process [1–3]. The following are a variety of processing methods within the open and closed moulding categories.

7.1 Open Mould Processing Technologies

Open moulding processes are those where the part being manufactured is exposed to the atmosphere. The composite part is typically handled manually. The open moulding process involves saturating the fibre reinforcement with resin using manual roll-out techniques to consolidate the laminate and remove any entrapped air. In the open moulding process, no closed mould or external pressure is applied. A major factor in this operation is the transfer of resin from a drum or storage tank to the mould. The resin mixture may be a liquid being formed onto a reinforcing material, or it may be in the form of a prepreg material being formed for the final cure [4, 5].

7.1.1 Contact Layup

- *Processing Technology*—Contact layup (also known as hand layup or wet layup) is one of the earliest processing techniques used for the fabrication of composites. In this batch process technology, reinforcement in the form of fabric or chopped strand mat (CSM) is used, and the matrices are room-temperature curable thermosetting resins, such as polyester resins and epoxy resins. An open mould is used to fabricate the product, and the outer surface of the composite product is governed by the finish of the inner surface of the mould, as shown in Fig. 7.1.

  The open moulds are usually made out of wood, plaster of paris, composites, or steel, depending on the durability of the mould required. During moulding, first a release agent (such as wax polish or a PVA solution) is applied. Then a layer of high-viscosity resin, which is called *gel coat*, is applied on the mould in order to achieve the glossy finish of the product. Room-temperature (RT) curable resin is applied on every layer of preform sequentially. Layer after layer of preform is placed over the mould until the desired thickness of composite is achieved. The resin is applied using a brush and a roller. After the curing is complete, the product is ummoulded. This method is simple and easy to handle, but it is labour-intensive and not amenable to high production rates. This process does provide a good surface finish on one surface (i.e. moulding surface). The drawback of this technology is that the quality of

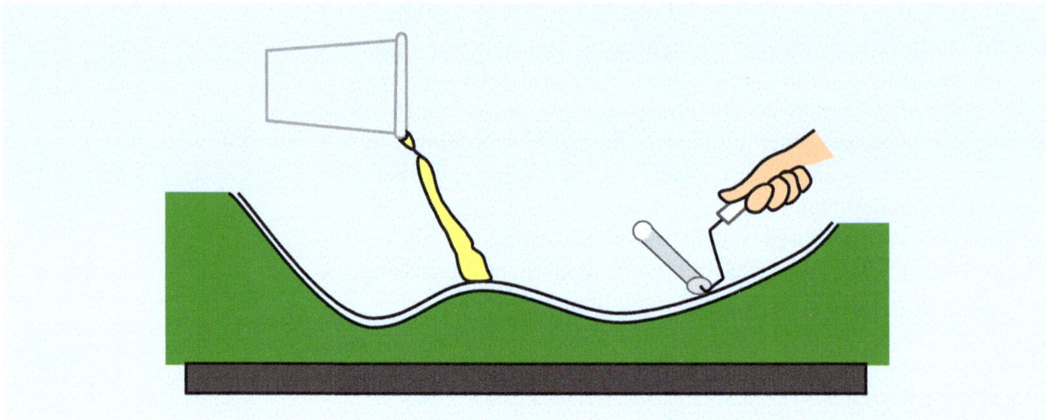

Fig. 7.1 Contact layup process [6]

7

the composite is not very good because many flaws (voids, uneven distribution of reinforcement and matrix, uneven thickness, etc.) are present. The biggest advantage of this process is the low cost of the mould with respect to the size of the composite produced. Contact layup is used for the manufacture of large products where moulding in a closed mould is not feasible or cost-effective. The composite products made by this technology need to be designed with high factor of safety (FoS) due to the high possibility of flaws in the composite. A new variant of this technology is the use of pre-impregnated reinforcement (prepregs) where the layup is carried out with the use of semiautomatic rolling and pressing systems. The quality of the product obtained is good, and the FoS can be kept lower in comparison to the contact layup process carried out with a manual impregnation process [7].

Point to Ponder…

W*indmills are the second most important source of non conventional energy after solar energy. The potential of wind energy from in India and the world is 50,000 MW and 500 GW, respectively.*

Contact layup technology is used for making boat hulls, windmill blades, rail coaches, containers, front and rear spoilers for automotives, etc. A blade for a windmill is made using the manual or hand layup process with an open mould and raw materials in the form of prepregs. The world's largest windmill rotor blade (B75) made by Siemens is shown in Fig. 7.2a, and the picture of a moulded blade is given in Fig. 7.2b. None other than this process—however automated it may be—can replace this method for the production of a composite blade of such a large size, usually >15 metres in length, for a windmill.

- *Processability Parameters*—The following are the processability parameters for making polymeric composites using a contact layup process.
    - *Viscosity of the resin*—Low viscosity of the resin facilitates the processability of the composite in a contact layup technology. The viscosity of the resin needs to be the minimum to enable it to easily percolate into and completely wet the reinforcement. High viscosities of resin do not favour the processability of composites using contact layup technology.

Fig. 7.2 **a** B75 windmill rotor blades by Siemens and **b** moulded windmill rotor blade [8, 9]

- *Gelation time of resin*—The gelation time controls the processability of a composite because it provides the processing window in terms of time. The gelation time must be sufficiently long so that the reinforcement can be placed on the mould, reinforcement can be impregnated with the resin, and the next layer of reinforcement can be placed. The relationship between gelation time and wettability time is given here.

$$T_G \propto \frac{1}{\eta} \quad (7.1)$$

$$T_W \propto \frac{1}{\eta} \quad (7.2)$$

$$T_W \propto T_G \quad (7.3)$$

$$T_W = M\, T_G \quad (7.4)$$

where:

η = Viscosity of the resin
T_W = Wettability time for the reinforcement
T_G = Gelation time of the resin
M = Processability constant

- *Form of reinforcement*—Only fabrics and chopped strand mats (CSM) are processable by contact layup technology. The processability of composites having a curvilinear or 3-D profile is governed by the ability to drape the reinforcement on the contoured surface. The drapeability of CSM and satin weave fabrics is the best, while for plain weave fabrics, it is the least. The descending order of drapeability of the reinforcement is CSM → eight-harness satin weave fabric → six-harness satin weave fabric → four-harness satin weave or twill weave fabric → plain weave. The satin weave of a harness more than eight is difficult to handle; therefore, only up to an eight harness satin weave fabric is used. The drapeability of CSM is the best because there is no interlacing point, and strands of fibres (e.g. glass fibres) are loosely bonded. Therefore, CMS provides the maximum pliability leading to a maximum drapeability [10]. On the contrary, plain weave has interlacing points in close proximity; therefore, it is not pliable but is stiff and not drapeable.
- *Shape and profile of the product*—Sharp edges and undercuts do not favour the processability of composites by contact layup technology. Albeit, undercuts in the product can be processed by using split moulds, but this makes the moulding process complicated and costly. Contact layup offers a good finish on only the surface of the composite, which is in contact with the mould.

7.1.2 Spray Layup

- *Processing Technology*—The spray layup (spray-up) process is also called the chopping process and is similar to the hand layup process. The spray-up process is used for making boats, storage tanks, transportation containers, bathtubs, bathroom fittings, etc. in a large variety of shapes and sizes. A laminate with chopped fibres has good conformability, and its processing is faster than hand layup in moulding complex shapes.

  Similar to contact layup technology, in this process, RT curable resins (such as epoxy or polyester) only are used. In the spray-up process, the thickness and consistency is controlled manually; therefore, this process is more operator-dependent than compared to hand layup. Although the production rate per mould is low in the spray-up process, the consistency in quality of the composite product is good. However, it also is feasible to produce substantial production quantities using multiple moulds. In this process, a gel coat is first applied on the mould prior to the spray-up of the laminate system. Continuous fibres, glass rovings, and resin are mixed with hardener and are fed through a chopper gun. The resin-saturated fibre strands on the mould are shown in Fig. 7.3. Then the compacting rollers are applied on the laminate sprayed on the mould to saturate the fibres with resin and to make the

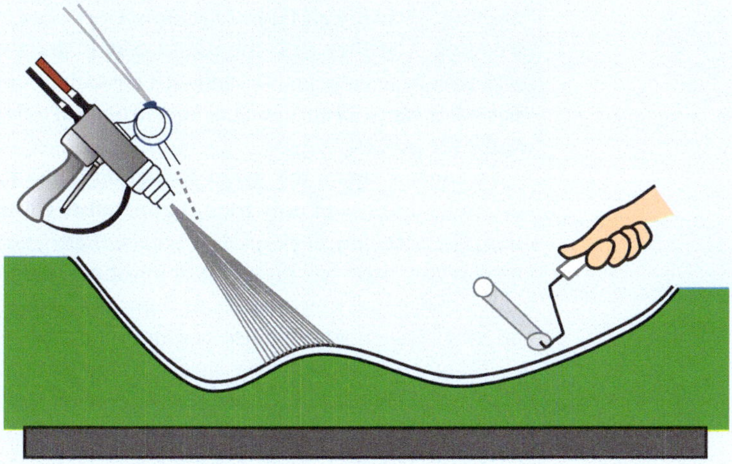

Fig. 7.3 Spray layup processes [6]

laminate more compact and devoid of air gaps. Additional layers of chopped laminate are sprayed to achieve the required thickness.

- *Processability Parameters*—Similar to contact layup, this process has the following processability parameters.
  - *Viscosity of the resin (η)*—The viscosity of the resin needs to be low so that resin can be sprayed through nozzle. The viscosity of the resin is preferred in the range of 500–5000 cps for this process. In order to reduce the viscosity of the resin, diluents may be added in order to improve the processability of the resin.
  - *Gelation time*—The gelation time of the resin needs to be high in order to provide the processing window for the resin and fibres to be placed in the required position on the mould.
  - *Form of reinforcement*—In this process, only staples of fibres are used as reinforcement. The length of the fibres is controlled during the layup process. The processability of reinforcement also is restricted to staples of fibres. This process does not offer processability for reinforcing fabrics and particles.
  - *Shape and profile of the product*—Sharp edges and undercuts do not support the processability of composites using spray layup technology. Although undercuts in the product can be processed using split moulds, this makes the process complicated and costly. Spray layup offers a good finish on only the surface of the composite that is in contact of the mould.

Contact layup and spray moulding are by far the most common processes used to fabricate composite parts, accounting for over 40% of composites processed worldwide. These are relatively simple processes with low investment costs but a high degree of manual handling. Virtually all types of reinforcement can be used in open moulding. Together with the use of core materials, the reinforcement may be used to create sandwich structures, enabling access to the widest range of mechanical and structural performance of any composite process.

In contact layup, the major limitation is the low-dimensional accuracies and use of only room-temperature curable resins. More often, the mechanical properties obtained are poor. Thus, contact layup is not used for high-end applications requiring more of a load. Unsaturated polyester resins dominate in this area, but epoxy and vinyl ester resins are also common for load-bearing composites. Spray lamination is very similar to the hand layup process, as discussed previously. This lamination process is much easier and faster than a hand layup and can be automated using automatic robotic spray guns. The major limitation is the control of volume fractions of the reinforcement where usually the volume fraction of the reinforcement remains at the lower side.

7.1.3 Filament Winding Technology

Point to Ponder…
Geodesic path is a shortest path between two vertices. Generally, filament winding is carried out on a geodesic path. Non-geodesic path filament winding requires high tackiness of the resin and special processing features for fibre tension and holding.

- *Processing Technology*—Filament winding is a process which is used for fabrication of cylindrical composite structure for high-pressure applications. It is an automated open moulding process that uses a rotating mandrel as the mould. The filament winding process is mostly used for thermosetting resins. In this process, roving of glass or aramid fibres or tows of carbon fibres are first impregnated in the resin bath and then passed through the nip roller in order to remove the excess resin. This impregnated roving or tow is wound over a mandrel under a specific tension on a geodesic path. The geodesic path is a firm path on a curved surface on which no lateral movement or slippage of fibre rovings or tows is allowed [11, 12]. A schematic of the filament winding process is shown in Fig. 7.4. The laying of the roving or tow is controlled by a three to ten axes of the machine. The most important axes of control are movement of the carriage; movement of the payout eye, which lays the roving on the mandrel; rotation of the mandrel; movement of the nip rollers; and tension in the roving/tow.

  The rocket motor of a spacecraft only can be made using filament winding or polar winding technology out of carbon or Kevlar fibres as the reinforcement and epoxy or phenolic resin as the matrix systems. The filament winding process is found most suitable for making composite pressure vessels, such as a rocket motor. However, a variant of the filament winding process is the tape winding process, which is presently used for the manufacturing of large-sized pressure vessels due to its fast processing time.

  In the *filament winding process*, the mandrel acts as a male mould. This produces a finished inner surface and a laminated rough surface in composites. Filament winding results in a

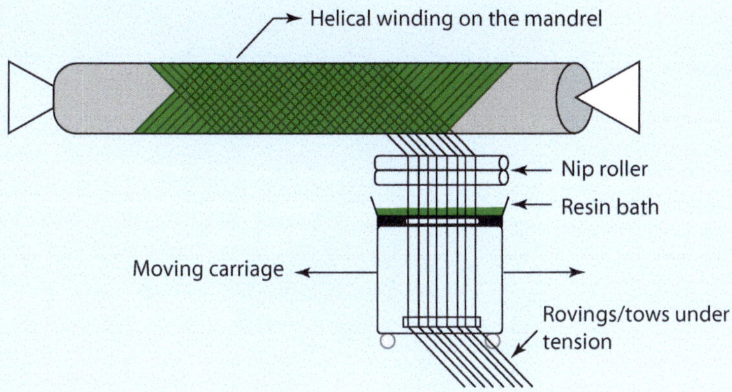

Fig. 7.4 Schematic of filament winding process

high degree of fibre loading, which in turn provides high mechanical properties, such as the tensile strengths and Young's modulus. The filament winding process is generally used for fabrication of tubular structures, such as cylindrical products like chemical and fuel storage tanks, pipes, stacks, pressure vessels, and rocket motor cases. Mandrels of suitable size and shape made of steel or aluminium form the inner surface of the hollow part. Some mandrels are collapsible to facilitate part removal.

In this process, a release agent is applied over the mandrel. The mandrel has minor tapper so that after curing the moulding can be removed easily by sliding. If product is big, collapsible mandrel are used to extract the moulding. Filament winding has two types of windings: helical and hoop. *Helical winding* is carried out at a certain angle other than 90°, while *hoop winding* is carried out at a 90° angle to the mandrel. Hoop winding is used for containment of very high pressures. Helical winding provides structural integrity to the filament wound structure. The combination of helical and hoop winding gives the correct combination of properties to a composite pressure vessel.

In order to process thermoplastic filament wound composites, thermoplastic towpregs (pre-impregnated fibre tow and roving) are used with special add-on heating and pressing roller mechanisms in the filament winding machine. The towpregs for the filament winding operation can be made using hot melt coating process [13]. At the temperature above the melting temperature, the viscosity of thermoplastics is reduced, and this allows for better fibre coating. High-performance thermoplastics are also used in this process. A filament wound glass fibre-reinforced polyether ether ketone (PEEK) composite pipe is shown in Fig. 7.5. This pipe is used in oil and gas industries because PEEK has good oil and gas resistance. The hydrolytic stability of PEEK is also good, and its moisture absorption is less than 0.2%.

Fig. 7.5 Filament wound glass fibre-reinforced PEEK pipe [14]

- *Processability Parameters*—The following are the processability parameters for filament winding technology:
  - *Winding path*—In this technology, the filament can be wound on a geodesic path only. A non-geodesic path will not allow the wounded filaments to remain at that position; rather, they will slip away. Non-geodesic path winding is an area of current research and interest. In this process, it is necessary to have raw materials that provide high friction so that the filament wound fibres do not slip from their position. Also, the winding path is designed in such a way that the wound fibres get physically anchored by the previously wound fibres.
  - *Tenacity of fibres*—In filament winding technology, only high-tenacity fibres are processable. The endurance of rovings or tows must be high enough to sustain the tensile force of winding.
  - *Form of reinforcement*—In filament winding technology, only reinforcement in the form of rovings or tows is processable. This technology also does not allow the processability of fabrics and particulate reinforcements.
  - *Fuzzing of fibres*—Poor surface treatment, poor storage conditions, and breakage in the fibres cancause fuzzing (collection of fibres) at the payout eye. This phenomenon causes a reduction in the bandwidth of the wound fibre on the mandrel, causing a reduction in the strength of the composite. Fuzzing is mostly observed in carbon tows. The layered structure of carbon fibre and its skin core heterogeneity causes fuzzing from the removal of the surface layer of carbon fibre.

- *Viscosity of the resin (η)*—The viscosity of the resin needs to be low so that resin is able to wet all filaments of roving and tow passing through the resin bath. The high viscosity of resin can adversely affect the processability of composites using filament winding technology.
- *Residence time*—The residence time of roving or tows is an important processability parameter. In this process, the residence time must be long enough so that all of the filaments of roving and tows are wetted with resin.
- *Winding speed*—The winding speed is an important processing parameter that indirectly controls the residence time of roving or tows in the resin bath. Low winding speeds cause less fuzzing of carbon fibres; thus, an optimum winding speed improves processability of the composite using filament winding technology.

Example 7.1 suggest the processing technology to be used for the fabrication of (a) a high-pressure vessel, (b) frame of a desert cooler, (c) a golf shaft, and (d) roof panels.

Answer

(a) *High-pressure vessel*—A small-sized, high-pressure vessel may be fabricated using a filament winding process, and a large-sized, high-pressure vessel may be fabricated using a tape layup process. Hoop winding would be needed in combination with helical winding based on the design requirements for the high-pressure vessel.

(b) *Frame of desert cooler*—This may be fabricated using a hand layup process with a wooden mould.

(c) *Golf shaft*—This may be fabricated using a filament winding process. In this process, predominantly helical winding will be carried out.

(d) *Roof panels*—These may be fabricated using a hand layup process with either wooden or plaster of paris moulds or patterns.

7.1.4 Polar Winding Technology

In polar winding technology, the winding takes place in a circular path (i.e. pole to pole). This allows both geodesic and non-geodesic filament winding. In this process, the mandrel is supported by one end, which facilitates the long helical winding from pole to pole. The fabrication of a polar wound composite product is shown in Fig. 7.6.

The fibre delivery system (payout eye) in this technology can be moveable or stationary. In a *stationary system*, mandrel rotates end to end in addition to rotating on its polar axis. A *movable delivery system* rotates around the mandrel in a polar orbit, and the mandrel rotation is confined to incremental rotation. Polar winding technology provides the possibilities for non-geodesic path winding with specially designed mandrels. Care must be taken for the sagging of the mandrel

Fig. 7.6 Polar winding technology [15]

when a moving delivery system is used. The bend in the mandrel may affect the composite product. In a stationary delivery system, the whipping action of the mandrel must be avoided to maintain a good quality of the composite product. Mounting the mandrel in a vertical postion avoids this bending problem. All of the processability parameters applicable to filament winding technology are applied in this technology. Polar wound CNG storage tanks are being used in automobiles, and these tanks are capable of taking 200–250 bar of pressure.

7.1.5 Automated Tape Winding Technology

Point to Ponder…
R*ocket motor casings are made from an automated tape winding process. The temperature and pressure experienced in rocket motor casings are to the tune of 2000 °C and 1000 MPa, respectively.*

- *Processing Technology*—This process is similar to the filament winding process except in this process, in place of filaments, prepreg tape is used. The prepreg tape material is fed through an automated tape application machine (robot). The prepreg tape is wound over a mandrel in a specific designed pattern for helical and hoop winding. Due to the thickness of the tape near a non-geodesic path, winding is also possible using a tape winding process. In a tape winding process, multiple layers of prepreg tape also are applied to achieve the design thickness. The tape width decides the winding patch. The thickness of the tape decides the number of winding layers required. A cassette tape layer also has been introduced, but this is not in widespread use. This method was introduced to provide quick fabrication of composite part in tape winding process. An example of an automated tape and filament winding process is shown in Fig. 7.7.

  The tape laying process has emanated as an evolutionary—not a revolutionary—process. In very large composite products, filament winding takes a very long time to fabricate the product because the number of passes required to achieve the required thickness is very high due to the small winding width

Fig. 7.7 Automated tape and filament placement [16]

(approximately 5 mm) offered by rovings and tows after being spread on the mandrel. However, the number of passes required by tape having a width of approximately 100 mm is much less than that of roving and tow winding. Thus, the processing is reduced. The tape laying of thermoplastics also has been tried; however, the overall interest in this process dwindled due to complexities of processing and the high-energy requirement that did not measure up to the viable economies of production. Most tape laying of thermoplastics involves hot shoes or laser welding. Some work on PEEK has been carried out using an amorphous film carrier, which becomes molten material when heated. This severely limits the temperature capability of the system, but it does not provide good laminates without subsequent autoclave compaction.

- *Processability Parameters*—The following are the processability parameters for automated tape winding technology.
  - *Winding path*—In this technology, the tape can be wound on a geodesic path only. A non-geodesic path will not allow the wound tape to remain at a required position; rather, they will slip away.
  - *Tensile strength of tape*—In tape winding technology, tape is wound under tension. Therefore, it is essential that the tape must have high enough tensile strength to sustain the force of winding.
  - *Tackiness*—Tackiness is an important processability parameter in tape winding technology. Low tackiness of prepregged tape does not allow it to remain on its wound position; therefore, the quality of composite deteriorates. Only good tackiness allows the processability of prepregged tape in a tape winding process. The tackiness is ensured by the 'B' staging of the matrix resin in the tape prepreg and its proper storage thereafter.

7.1.6 Vacuum Infusion (VI) Technology

- *Processing Technology*—The *vacuum infusion* process is akin to vacuum bagging. In this process, reinforcement is placed between the mould and the vacuum bag, and then the vacuum is applied. While under the vacuum, the resin is applied on the reinforcement so that the reinforcement is impregnated. Due to the vacuum process, the resin is sucked very efficiently inside the vacuum bag and penetrates all over the reinforcement. The pressure in vacuum infusion technology is kept at 0.5–10 bars inside the vacuum bag.

  Open moulds are used in this process. The reinforcement and core material are laid to dry in the mould. This is done manually with precise positioning of the reinforcement. When the resin is sucked into the mould, the laminate is already compacted; therefore, there is no room for excess resin. Excellent volume fraction of the reinforcement is possible with vacuum infusion. Therefore, mechanical properties of the laminate are excellent. Vacuum infusion is suitable to mould very large structures, but the production rate in this process is low.

  In order to achieve glossy outer surfaces, a mould may be gel coated. After curing of the gel coat, the dry reinforcement is placed in the mould. The reinforcement includes a reinforcing fabric or mat and core material if required. A perforated release film is placed over the dry reinforcement. A mesh or a 'crinkle' ply is positioned to facilitate the distribution of the resin across the reinforcement. The vacuum bag is then positioned and sealed at the edges of the mould. The vacuum bag is connected to the resin container. A vacuum is applied to compact the resin for cementing, and the resin is sucked into the mould, as shown in Fig. 7.8.

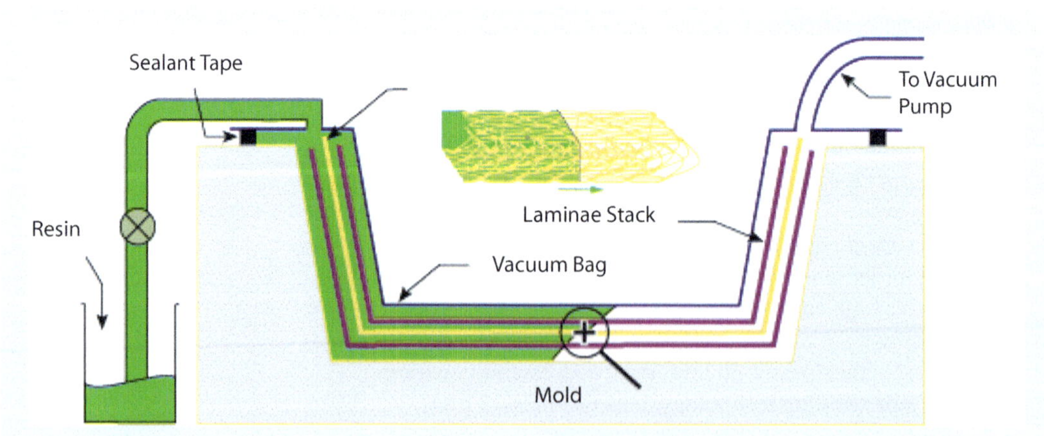

Fig. 7.8 Picture of vacuum infusion processing [17]

Vacuum infusion produces composites with a uniform resin distribution, producing high-quality composite structures. The tool cost is low in this process because open moulds are used. In vacuum infusion technology, there is no residue or precipitates from substantial emission reduction compared to other open moulding or vacuum bagging processes. In Brazil, SkyRail coaches have been made using VI technology [18]. The construction of these coaches uses sandwich composites with a foam core. These rail coaches are capable of ferrying 10,000 passengers per day at a speed of 65 Km/hr.

- *Processability Parameters*—The processability parameters for vacuum infusion technology are the following.
  - *Viscosity of the resin* (η)—Only low viscosities of resin allow the processability of composites. Low viscosity is essential for the flow of resin through compacted reinforcement for complete wetting.
  - *Gelation time*—Gelation time for the resin needs to be higher than the impregnation time for the processability of the composite.
  - *Shape and profile of the product*—Sharp edges and undercuts do not favour the processability of composites by vacuum infusion technology because sharp edges may puncture the vacuum bag.
  - *Form of reinforcement*—Only fabrics and CSM are used as reinforcements in order to allow the processability of the composite using vacuum infusion technology.
  - *Cure temperature*—In general, the matrix resin used in vacuum infusion cures at room temperature. This process does not support high-temperature resin systems. For high-temperature curing resins, the cost of processing goes up due to the use of both high-temperature bagging film and high-temperature putty. There also is a risk of warpages and dimensional inaccuracies due to open moulding processing.

One disadvantage of VI is that the excellent consolidation favours a thin cross section with a high-fibre content. Thus, the composites developed may lack sufficient stiffness. In order to combat this problem, core materials of various types are used to increase the stiffness. A further complication with VI is that while the lamination process itself is very quick, the setting up of the mould with the bag and resin distribution network is time-consuming and costly, as the materials are disposable.

Typically, vacuum infusion is used for parts where high performance and quality are the main criteria and where the extra cost associated with the disposable materials can be supported. The process has been used particularly in the marine industry for boat hulls and masts.

7.1.7 Thermoforming

- *Processing Technology*—Thermoforming in composites processing is a secondary process where a thermoplastic composite sheet initially is fabricated using an extrusion, matched die moulding, or compression moulding process and then is thermoformed on a mould with or without using the vacuum. The main constraint in the thermoforming of thermoplastic composites is that only short fibre composites can be thermoformed.
- *Processability Parameters*—The processability in thermoforming is only for short fibre-reinforced semicrystalline or amorphous thermoplastic sheets because they have a softening temperature—not a sharp melting point as in the case of crystalline polymers. The processability of fabric-reinforced sheets is not feasible due to the restriction in stretchability caused by the fabric.

7.1.8 Open Casting Process

- *Processing Technology*—In this casting process, reinforcement and resin are mixed and cast in a open mould. The disadvantage of the open casting process is that the volume fraction of reinforcement is quite low and a true composite is never achieved. The viscosity plays an important role in open casting, and this process is most suited for thermosetting resin systems, such as epoxy or polyester.
- *Processability Parameters*—All of the processability parameters discussed in spray layup technology are operable here. The open casting process also supports particulate reinforcement.

7.1.9 Prepreg Processing

The term *prepreg* is actually a short form of *pre*-im*preg*nation. A prepreg is a reinforcement pre-impregnated with resin. Prepregging is the process in which the reinforcement is impregnated with the resin and a curing agent mixture. The curing agent is mixed with a retarder that, at a low or ambient temperature, does not cause cross-linking of the matrix. The cross-linking of the matrix in presence of the retarder only starts at a high temperature. Dicyanamide (DICY) is commonly used as a retardant to the chemical cross-linking of epoxy resin in preparation of prepregs. These pre-impregnated reinforcements (also known as prepregs) are made using reinforcements in the form of woven fabrics, roving, and unidirectional tape. Pre-impregnated short fibres in the form of a sheet are called sheet moulding compounds (SMC), and in the form of dough, they are called dough moulding compounds (DMC). SMC and DMC are also classified as prepregs. In prepregs, fabrics and tapes are provided as continuous rolls in widths up to 2 m and lengths up to several hundred metres. The thickness of the fabric or tape constitutes one ply in the construction of a multi-ply layup. Impregnated roving is wound onto cores or bobbins and is used for filament winding. Once the resin mixture has been impregnated onto the fibres, the prepreg must be stored in a refrigerator or freezer. The cold storage prevents the chemical reaction from occurring prematurely and causing cross-linking of the matrix. Prepreg materials are used widely in advanced composite industries, particularly in the aerospace industry.

The term *preform* is short for *pre*fabricated *form*. Preforms are made by braiding, weaving, and/or cutting and stitching the reinforcing fabric into the desired shape. Preforms are the pre-shaped reinforcement that ensures the firm, positional placement of the reinforcement and avoids the placement of the reinforcement layer by layer. The placement of reinforcement layer by layer is time-consuming and delays the total processing cycle. Preforms in a desired shape and size may be prepared separately. After impregnation and curing, the preform will result into a final product. Preforms enhance the processability of polymeric composites. Preforms are often used in a resin transfer moulding (RTM) process. Preform-based moulding is required where high-dimensional accuracies and consistency in the properties of polymeric composite are needed. Glass braiding-reinforced thermoplastic pipes are shown in ◘ Fig. 7.9.

Prepregs in composites are becoming increasingly common due to their ease of use, consistent properties, and high-quality surface finish. Different types of resins, including the majority of thermoset and thermoplastic matrices, are used in making prepregs. The processibility of polymeric composites is influenced by the properties of the prepreg. Prepregs limit the deviation in processing parameters and processability to a great extent. Many of the processability issues related to reinforcement and matrix in terms of their volume fractions are well controlled with prepregs.

◘ **Fig. 7.9** Glass braiding-reinforced thermoplastic pipes [19]

Preforms and prepregs are used for the processing of polymeric composites for excercising better control over the properties and dimensional control on the moulded product. They provide a flexible window for the processing of polymeric composites. The greatest advantage of using prepregs is their ease of use. If the polymeric composites are made by using liquid resin in a matched die moulding process, there would be a need to use fabric, matrix, and other additives for fabrication of the composite product. Most thermosetting matrix systems are considered hazardous when coupled with messy handling of liquid resins. Prepregs come in the form of dry or sticky roll and has the desired amount of both resin and hardener already impregnated in the fabric.

Most thermosetting prepregs come with a separation film on both sides of the fabric to protect it from dust and contamination, prevent the sticking of adjacent layers of prepreg in the roll, and make transportation easier. The prepreg is then cut to the desired shape, the separation film is peeled off, and the prepreg is then laid onto the mould for curing the polymeric composite part. The head of a golf club made out of carbon fibre prepreg is shown in ◘ Fig. 7.10.

The disadvantage of thermosetting matrix-based prepregs is that they have to be kept under refrigeration in order to prevent premature curing and to improve their shelf life. The processing of composites using prepreg involves two stages. First, the manufacturing of prepregs is carried out. Second, the manufacturing of the polymeric composite part is carried out using the prepregs. This increases the fabricating cost of the polymeric composite.

- *Thermosetting Prepregs*—Thermosetting prepregs are more commonly used in composites' industries. The most common resin matrix used in thermosetting prepegs is epoxy. However, other thermosetting resins, such as bismaleimide (BMI) and phenolic resins, are also used. The schematic of the prepregging process is shown in ◘ Fig. 7.11. Thermosetting prepegs are

◘ **Fig. 7.10** Head of a golf club manufactured by using carbon fibre prepreg [20]

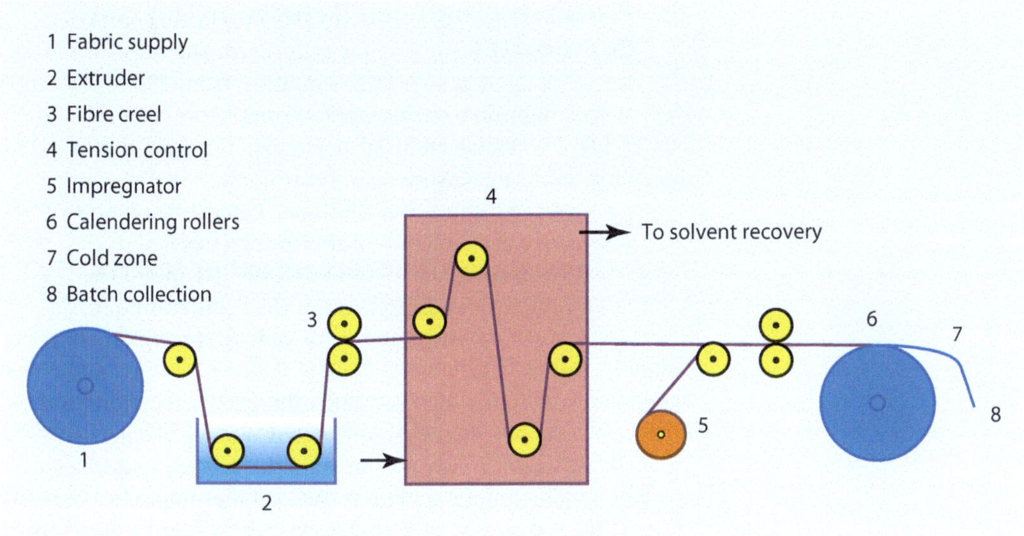

Fig. 7.11 Schematic of prepreg process [21]

sandwiched between two separating films (commonly low-density polyethylene film) so that the adjacent layers don't stick together during storage.

- *Thermoplastic Prepregs*—In thermoplastic prepregs, thermoplastic resins are used. Unlike thermosetting prepregs, thermoplastic prepregs do not have various stages and thickness. They are dry prepregs, and on the application of heat and pressure, they consolidate in the form of composites. Thermoplastic prepregs incorporating fibreglass, carbon fibre, aramid, etc. as reinforcement and PP, PET, PE, PP, PEEK, etc. as resin matrices are available.

The primary difference between thermosetting and thermoplastic prepregs is that thermoplastic prepregs are stable at room temperature and generally have a longer shelf life compared to thermosetting prepregs.

In thermosetting prepregs, the thermosetting resin is in a liquid state and is applied on the fabric or tape so that it fully impregnates the fibre reinforcement. Excess resin is precisely removed from the reinforcement. Meanwhile, the resin undergoes partial curing (i.e. changing the state of the resin from liquid to semisolid). This is known as the *B-stage*. In the B-stage, the resin is partially cured and usually feels tacky. When the resin is brought up to its curing temperature, it often returns briefly to a liquid state prior to cross-linking (hardening) completely. This helps in the percolation of the resin among the filaments of the reinforcement due to the low viscosity of the resin. Once cured, the thermoset resin that was in the B-stage gets fully cross-linked. This cross-linked stage is called the *C-stage*.

7

7.2 Processing-Induced Defects in Polymeric Composites

Defects in composites occur either during processing or in their service life. Processing-induced defects in composites may occur due to the following reasons.

7.2.1 Unprepared Reinforcement or Matrix

The surface of the reinforcement must be hydrophobic for the compatibility with a hydrophobic matrix. If the surface treatment for compatibility with the matrix is poor, the composite will be of poor quality, having low strength, low modulus, and low interlaminar shear strength (ILSS). Also, the surface treatments are responsible for efficient handling of the reinforcement. If the surface coating of the carbon fibre is defective, a fuzzing of carbon fibre is observed during the filament winding at the payout eye. Usually, a coating of polyimide or epoxy resin is added to the carbon-fibre surface to make it compatible with organic resins and to provide ease of handling.

As far as matrices are concerned, the mixing of all the components (in two- or three-component systems) must be mixed in correct proportions to have the optimum properties. A high content of hardner causes high exothermic reactions and a high cross-link density, leading to brittleness and ultimately to poor-quality composites. Similarly, a low hardner content leads to under curing or no curing of the composite, affecting its glass transition temperature and mechanical properties. In one component matrix system, such as phenol formaldehyde resin (popularly known as phenolic resin), lumps of resin due to ageing are observed that also cause poor composite quality. The processing-induced defects for open and closed mould processing are given next.

7.2.2 Improper Storage Conditions

Storage of resin is important because resins mature with heat and time. The maturity of resin is the initiation of a low degree of cross-linking that increases the viscosity of the resin. Resin matrices are susceptible to moisture absorption. The ingress of moisture causes the material to bulk and creates surface defects from high-temperature curing due to the formation of water vapours. The accumulation of dust and foreign particles on the surface of the reinforcement or inside the matrix reduces all of the properties of composites.

7.2.3 Wrong Placement of Fibres

Tailoring the in-plane stiffness of composite laminates can improve the buckling and post-buckling characteristics by redistributing the applied loads. The fibre placement technique is the core technology providing support to variable stiffness composites. The mismatch between designed and real fibre paths increases as the width of a single tow and the number of tows laid at the same time increase. This causes nonuniformity of the fibre orientation and some defects, such as tow gap and overlap and irregularity of the thickness distribution, that occur in composites.

7.2.4 Improper Volume Fractions

Improper volume fractions lead to either resin-rich areas or resin-starved areas. Both of these situations are detrimental to the performance of composites. For the best performance of polymeric composites, equal fractions of volumes of reinforcement and matrix must be kept.

7.2.5 Uncontrolled Temperature, Pressure, and Time

One of the major challenges in manufacturing of composites is the efficient handling of the fibres and the application of uniform pressure during the curing of the resin. The fibres are stiff in nature and do not flow. For this reason, continuous fibres are either stacked in layers that are successively impregnated with the resin or formed into a preform that is placed in a mould where the resin is supplied under pressure to wet the fibres. In addition, getting a high-volume fraction of fibres in the composites is a major concern. The properties of composites depend upon the appropriate curing of the matrix. The pressure, temperature, component geometry, and distribution of pressure over the surface of the component are some of the deciding factors determining the quality and reliability of the manufactured component.

A temperature gradient through the thickness of the composite does occur during the curing process. The insulating character of composites restricts the heat transfer and increases the level of residual stresses that are formed in the composite. This increases the risk of fibre waviness and wrinkles, leading to a risk of defects in the composite. Thick laminates are subjected to combined effects of exotherm from curing, leading to high stresses in off-axis plies and increased risk of waviness and wrinkles in the plies (specially with complex shapes).

7.2.6 Improper Closing of the Mould

Improper closing of the mould leads to dimensional instability of the cured product. It also may lead to the resin oozing out upon curing. The oozing resin can lead to an imbalance in volume fractions of both the reinforcement and the matrix, causing the composite to have low mechanical properties.

7.2.7 Demoulding Temperature

Composite moulding (product) and trim must be taken out of the mould at room temperature or at a prescribed demoulding temperature. If these building products are removed from the mould at a higher temperature than the demoulding temperature, warpages may occur in the product due to the generation of residual stresses. Residual stresses are caused by the thermal nonequilibrium between the reinforcement and the matrix at a high temperature, even though curing is complete. The reason for thermal nonequilibrium is the difference in linear thermal expansion between the reinforcement and the matrix. This problem is observed most often in carbon fibre-reinforced composites because the conductivity of carbon is much higher than the conductivity of the matrix. Therefore, the coefficient of linear thermal expansion of carbon fibre is more than the matrix.

7.2.8 Post-fabrication Finishing Operations

In many composite products, there is a need to carry out post-fabrication finishing operations, such as surface painting, metalizing, and determining how the assembly fits. All of these operations demand care in handling so that there is no damage to the surface. Many a time, surface defects are not visible, but they remain in the form of microcracks or crazing. The microcracks or crazing occur in the matrix, and later on, with the application of stress, it propagates into the bulk composite, causing failure.

7.3 Future Composite Manufacturing Technologies

A major breakthrough in composite manufacturing technologies is not likely to occur in the foreseeable future. Most likely, there will be a series of improvements to existing manufacturing technologies, and manufacturing concepts already generated will be proven. New developments in open mould composite manufacturing are expected to occur in the foreseeable future and are described here.

7.3.1 Automatic Filament Winding

This is a mature manufacturing technique that has been in existence for a long time. Improvements in automation, speed, variable thickness, pad-up insertion, quality consistancy, flexibility in fibre orientation, control of resin and void content, and shapes other than cylinders will be seen. A combination of robotic and traditional filament winding systems (with seven to ten axes) is already available in crude form. If this combined system is perfected, it will be able to wind complex non-axisymmetric shapes, such as 'T' and elbow shapes. One of the most critical requirements for successful implementation of this method is controlling the tension of the deploying filament during the winding process. This critical problem may be quickly solved with the help of software.

7.3.2 Automatic Tape Layup (ATL) Machine

Significant progress has been made in ATL technology. Both speed and accuracy have increased tremendously when compared to earlier ATL processes. Advancements in computer technology (hardware and software) have influenced ATL. Along with improvements in speed and accuracy, the capability in size of the layup area has also increased. Although a new breakthrough is not expected to occur in ATL technology, improvements will be incremental but continuous.

7.3.3 Automatic Ply Cutting (APC) Machine

This technology has made significant progress in recent years. Three different methods of cutting are used for an APC machine: mechanical, laser, and water. Each has its own advantages and disadvantages.

7.3.4 Tow Placement

Tow placement is a relatively new process and has received considerable attention in recent years. It combines the advantages of ATL and filament winding. Tow placement can fabricate complex-shaped structures without having limitations on fibre angles. It has the potential to reduce production costs significantly. This technology holds promise; however, its use at high production rates still remains to be seen. Future developments include optimized control systems, head position feedback, and in-process inspection for fast, accurate, and high-quality parts production.

To Be Remembered

1. Open moulding processes are those where the part being manufactured is exposed to the atmosphere.
2. Room-temperature curable resins are used in contact layup and spray layup technologies.
3. Gelation time of the resin needs to be more than the layup time for processing composites.
4. Generally, filament winding is carried out on a geodesic path.
5. Geodesic path is the shortest path between two points on a curved surface.
6. Excessive hardner content causes high exothermic reactions and high cross-link density, leading to brittleness and creating composites of poor quality.

7. Low hardner content leads to under curing or no curing of the composite, affecting its glass transition temperature and mechanical properties.
8. Improper curing causes residual stresses in the composites.
9. Core materials are used in composite manufacturing in order to increase thickness. This is a common practice in vacuum infusion technology.
10. Thermoforming is predominantly a process for short fibre-reinforced thermoplastic composites.
11. Prepreg is a reinforcement pre-impregnated with resin.
12. Preforms are pre-shaped reinforcement that ensures the firm positional placement of reinforcement.

Exercises

1. The viscosity of polyester resin (processability constant = 0.5) goes up from 2000 to 70,000 cps in 30 minutes at room temperature. How much time will this resin take to wet and percolate through the reinforcement to get a good composite?
2. Why does fuzzing takes place in carbon fibres and not in glass fibres?
3. How does the tension in glass roving affect the filament winding process? Why will high tension cause a resin-starved filament wound structure? Explain.
4. What makes a contact layup process most popular for composites?
5. Calculate the wettability time for glass fibre-reinforced polyester composites. The gelation time for polyester resin is 5 minutes, and the processability factor of polyester resin is 0.5.
6. The viscosity of resin reaches 50,000 cps in 20 minutes at 100 °C during the processing of a composite. What will be the wettability time for fibres in this composite? (Processability constant is 0.5)
7. Outline the filament winding process for thermosetting and thermoplastic composites. Also, compare the major difference in their processing.

References

1. Murphy, J., *The Reinforced Plastics Handbook*, Elsevier Advanced Technology, Oxford, 1994.
2. Bader, S., *Crystic Polyester Handbook*, Scott Bader Co., Ltd, Wellingborough, 1980.
3. Monk, J.F., *Thermosetting Plastics - Practical Moulding Technology*, George Goldwin, London, 1981.
4. Kia, Hamid G., "Sheet Moulding Compound," *Science and Technology*, Hanser, USA, 1993.
5. *The Use of Epoxy, Polyester and Similar Reactive Polymers in Construction*, Vols. 1, 2, and 3, CIRIA, London, 2000.
6. *Open Moulding*, Moulded Fiber Glass Companies, http://www.moldedfiberglass.com/processes/open-molding-process.
7. Sigg, A., De Oliveira, R., Hochman, L., Cazes, T., and Michaud, V., "Process induced defects in thick composite laminates," *Laboratoire de Technologie*

des Composites etPolymères (LTC), Ecole Polytechnique Fédérale de Lausanne (EPFL). MXH143, Station 12, CH 1015 Lausanne, Switzerland
8. Holloway, J., *Siemens unveils world's largest wind turbine blades*, http://www.gizmag.com/worlds-largest-wind-turbine-blades/23578/, *2012*.
9. *The Long and Windy Road: A Visit to the Wild Horse Wind Farm*, 15 May 2010, http://redmondlibrary.blogspot.in/2010/05/long-and-windy-road-visit-to-wild-horse.html, 2010.
10. Lee, S. M., *Handbook of Composite Reinforcements*, Wiley VCH, USA, Polar Winding, 1993.
11. *Filament wound products*, Kineco, http://www.kinecogroup.com/product_listing.php?catid=4, 2010.
12. "Innovation Composite Winding Technology for CNG Storage in the Automobiles Industry," *FRP Today*, 14 (11), November 2014, p. 16.
13. Rajpurohit, A., "Fiber Reinforced Composites: Advances in Manufacturing Techniques," *FRP Today*, 14 (04), April 2014, p. 22.
14. ".Victrex PEEK Polymer based Composites Enable High Performance in Extreme Oil & Gas Environments," *FRP Today*, 14 (04), April 2014, p. 30.
15. *Filament wound products*, Kineco World of Composites, http://www.kinecogroup.com/product_listing.php?catid=4.
16. *Continuous Filament Winding Machnine*, https://en.wikipedia.org/wiki/Continuous_filament_winding_machine
17. *Organizing your composite workshop*, HomMaCom, http://www.composites.ugent.be/home_made_composites/organizing_your_composite_workshop.html.
18. "Brazil installs 100% composite air train in preparation for world cup," *FRP Today News*, 14 (1), January 2014, p. 22.
19. *Hose*, https://en.wikipedia.org/wiki/Hose.
20. *Club crown*, Three Guys Golf, http://www.threeguysgolfblog.com/club-crown/.
21. *Advanced Thermoplastic Composites*, AZO Materials, https://www.azom.com/article.aspx?ArticleID=319.

Processability in Closed Mould Processing of Polymeric Composites

A. K. Ghosh, M. Dwivedi, *Processability of Polymeric Composites*,
https://doi.org/10.1007/978-81-322-3933-8_8

The flexibility in composite processing provides possibilities for particular types of composites using different processing technologies, depending on the product requirement. The processing of composites and the process compatibility of composites are interlinked. *Process compatibility* is referred to in terms of economics of production, rate of production, processing technology, selection of raw materials, processing parameters, and post-fabrication processes. As a matter of fact, all of these parameters govern the processability of composites discussed in ▸ Chap. 4. The success of composite product design and production lies in the process adopted for its realization.

The processes in closed moulding are matched die moulding, compression moulding, injection moulding, pultrusion, resin transfer moulding (RTM), vacuum bag moulding, vacuum infusion processing, and autoclave moulding. In *closed moulding*, the composite is processed in either a two-part mould set or within a vacuum bag [1, 2]. A wide range of different processes have been developed for the moulding of composites parts, ranging from very simple manual processes (such as hand layup) to very sophisticated and highly industrialized processes (such as VARTM and autoclaving). Each process has its own benefits and limitations that make it applicable to particular applications. The choice of process is important in order to achieve the required technical performance at a reasonable cost. There are a variety of closed mould processing methods as described next.

8.1 Closed Mould Processing Technologies

Closed moulding methods are those in which all or part of the product is processed in a closed vessel or chamber. Liquid resin mixtures or prepregs are used for processing. In the case of liquid resin mixtures, they may be pumped into the mould for curing. In closed moulding, the composite part is manufactured in a closed mould, or it passes through a die to form the product. The heat and pressure then may be supplied in the closed moulding process. Some of the closed moulding processes are described here.

8.1.1 Matched Die Moulding

- *Processing Technology*—This moulding process is similar to contact layup except it uses a matched die mould that has a matching core and cavity, as shown in ◘ Fig. 8.1. In a matched die moulding process, there is a possibility of the use of a variety of resins (such as thermosets and thermoplastics) and reinforcements. The matched die mould ensures a good surface finish on both sides of the product due to intimate contact of the product on the core as well as the cavity of the mould. The matched die mould has the capabilty of heating and cooling the mould. Also, there is no limitation of the layup time in cases of high-temperature curing systems.

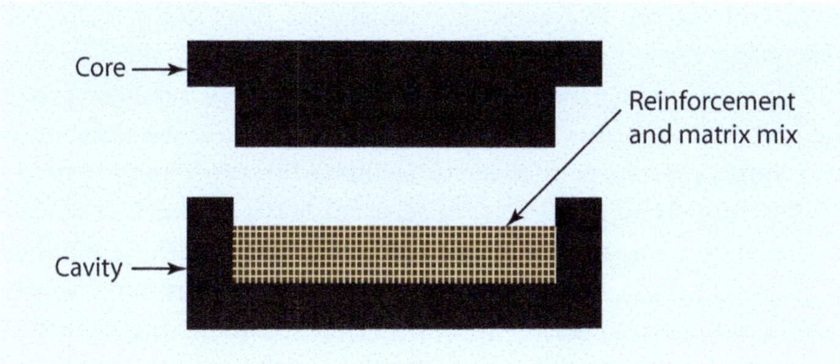

Fig. 8.1 Matched die moulding process

This process is also used for thermoplastic resins. In the matched die mould process, the reinforcement and resin mix is placed in the mould cavity. Then the core is clamped on the cavity with the help of fastening screws or with a hydraulic machine. The space between the core and the cavity decides the thickness of the product. The heating of the mould depends upon the matrix being used.

Example 8.1 Calculate the number of reinforcement layers in a glass-reinforced epoxy resin laminate with a thickness of 3 mm. The density of E-glass, area density of E-glass fabric, nominal thickness of E-glass fabric, and density of epoxy resin are 2.54 g/cm^3, 440 GSM, 0.33 mm, and 1.2 g/cm^3, respectively.

Answer

- Suppose (length × width) of laminate = 300 mm × 300 mm
- Thickness of laminate = 3 mm
- Volume of the moulding cavity = 270 cm^3
- Projected area of the mould cavity = 30 × 30 cm^2 = 900 cm^2
- Weight of one layer of glass fabric of dimension 30 × 30 cm^2 = 39.6 g

The ideal composite has 50% reinforcement and 50% matrix by volume. This results fall into 66% of glass fibre and 34% matrix by weight. Therefore,

- Theoretical density of composite laminate as per the law of mixture = 1.96 g/cm^3
- Weight of composite moulding = 270 × 1.96 = 529.2 g
- Weight of glass fabric = 349.3 g (at 66% weight %)
- Weight of epoxy resin matrix = 179.9 g (at 34% weight %)

Therefore,

- Number of glass fabric layers = 349.3/39.6 = 8.8 layers

Glass fabric layers can't be used in fraction parts. Therefore, eight layers will be used for the laminate. The fractional value will be avoided.

The theoretical calculation of Example 8.1 suggests that 8.8 (say nine) layers of glass fabric of composite laminas are required to achieve a composite that has equal volume fractions of reinforcement and matrix. However, this calculation does not account for processing errors, such as improper mould closure, dimensional inaccuracies during layup, and void content. These deficiencies are the practical problems, and during fabrication, these are difficult to overcome. Therefore, in order to facilitate the processing of composite of good quality, 01 layer of glass fabric (01 lamina) is reduced in the layup plan, and the composite is fabricated with 08 number of glass fabric (08 number of lamina) layers. This amounts to the removal of approximately 10% of the reinforcement. This argument drives an important point that the processability of a composite depends upon the right theoretical calculations as well as on the decision of striking balance among many factors that have been discussed in ► Chap. 3. This example cites the importance of striking a correct balance between the reinforcement and the matrix.

The matched die moulding process is economically the second-best processing method—after contact layup—for composites. This process provides good control over dimensional accuracy; therefore, it is widely used for products where dimensional tolerances are tight. The closing of the matched die mould can be controlled using hydraulics, screw pressure, or by tightening nuts and bolts, depending upon the size of the mould. The products made from this process are radomes, nose cones, protective covers, packaging, automotive parts, home appliances, electrical goods, etc.

8

8.1.1.1 Film-Stacking Process

Thermoplastic composites using continuous fibres are fabricated using a *film-stacking process*. This process involves a layered construction with alternate layers of thermoplastic films and reinforcement, as shown in ◘ Fig. 8.2.

When this stack is heated in the closed mould, a composite is formed. Advantages of film stacking over hot melt coating are the following:

- Processing temperature for film stacking is lower than hot melting.
- The matrix has to undergo one heat cycle compared to the hot melting process, where the first heat cycle is for prepregging and the second cycle is for moulding the product, allowing for fewer chances of thermal degradation.
- The film-stacking processing is more efficient as it forms a product in one processing cycle.

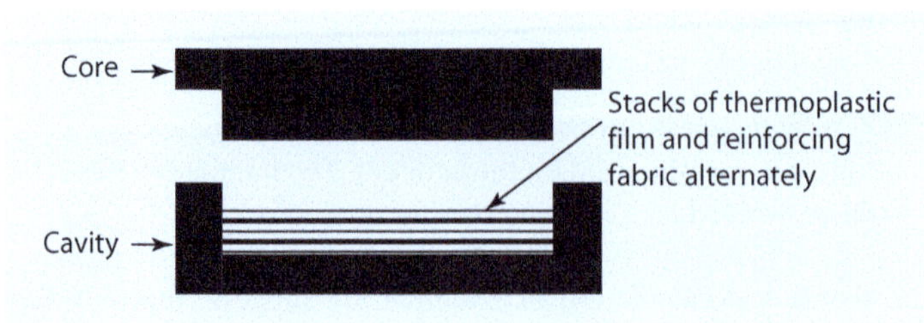

◘ **Fig. 8.2** Schematic of film-stacking process

- Film stacking is carried out at a temperature lower than the melting temperature, and the resin softens and flows under pressure, lowering chances of thermal degradation.
- Sandwich construction also may be used in this process.

The disadvantage of film stacking is that it requires more preparation time for placing the stacks correctly.

8.1.1.2 Commingled Fibres and Hybrid Fabrics

Another method of fabricating the thermoplastic composites using matched die moulding process is the moulding of commingled fibres or hybrid fabrics. In *comingled fibres* or *hybrid fabrics*, continuous fibres are spun using crystalline or semicrystalline thermoplastic resins. These fibres are interwoven with the reinforcement fibres at the individual fibre level. This causes the construction of a hybrid fibre bundle or fabric. This fabric can be heated above the softening or melting temperature of the thermoplastic resin so that the thermoplastic fibres melt, flow, and coat the reinforcement fibres [3–5]. This results in a thermoplastic composite. The potential advantages of this process are:

- The volume faction of the fibre can be precisely controlled.
- High-fibre content can be achieved.

8.1.1.3 Powder Impregnation Process

In yet another method, called the *powder impregnation process*, fibres are first passed through a chamber of charged and fluidized powder resin followed by passing them through an oven to make the polymer permanently adhere on the fibres. This technology can be used for the coating of rovings and tows, but good impregnation is difficult to achieve for reinforcement fabrics. The other method of impregnation involves fibre tows passing first through a fluidized bed of thermoplastic resin and then through the oven. The coated powder on the fibre is melted on the fibre surface in continuous operation, thereby forming a sheath of thermoplastic over the tow. The advantage of this processing technique is that a controlled deposition of the matrix over the reinforcement is achieved, providing good control over volume fractions of fibre and matrix. A disadvantage of this process is that coated powder (if not melted properly) gets separated from fibres while handling them [6–8].

8.1.1.4 In Situ Polymerization

In situ *polymerization* is another logical method of reducing the viscosity of the thermoplastic resin. This process utilizes monomers or solvated prepolymers of the thermoplastic to coat the fibres and then to polymerize the thermoplastic in situ. Only a few thermoplastic matrices (such as PEI and TPU) can be processed by this method. An advantage of this processing technique is that most difficult thermoplastics (PEI, PES, etc.) can be handled easily for moulding products. A disadvantage of this process is that the processing time is long even when compared to thermosetting resin-based composites [9, 10].

8

- *Processability Parameters*—Matched die mould process technologies allow the processability of all types and forms of reinforcement and matrices. These technologies provide the highest degree of freedom for processing composites. However, the following are the processability parameters for the various matched die mould technologies described previously.
  - *Temperature of the core and cavity of the mould*—A uniform temperature of the core and cavity of the mould is essential for the processability of polymeric composites using matched die mould technologies. The heat flow rate in both the core and the cavity must be controlled to have uniform temperature.
  - *Shape and profile of the product*—Sharp edges and undercuts do not favour the processability of composites by matched die moulding or compression moulding technologies.
  - *Melting or softening of matrix resin*—The processability of composites in these processes depends upon the softening of the matrix before curing. In the case of thermosetting composites, the melting of the resin matrix before the gelation is essential for the processability of the composite. In the case of thermoplastic composites, the softening or melting of the resin matrix is required for processability of composite. The processability of composites also depends upon an optimum hold time at the melting or softening temperature so that efficient percolation of the resin around the reinforcement takes place when the viscosity is the least.

8.1.2 Compression Moulding

- *Processing Technology*—Compression moulding is a batch process but can be modified into a semicontinuous or cyclic process, as shown in Fig. 8.3. This is a high-volume, high-pressure method suitable for the moulding of complex and fibre-reinforced plastic parts on a rapid cycle time. The compression moulding process also utilizes moulding compounds in the form of sheet moulding compound (SMC), bulk moulding compound (BMC), thick moulding compound

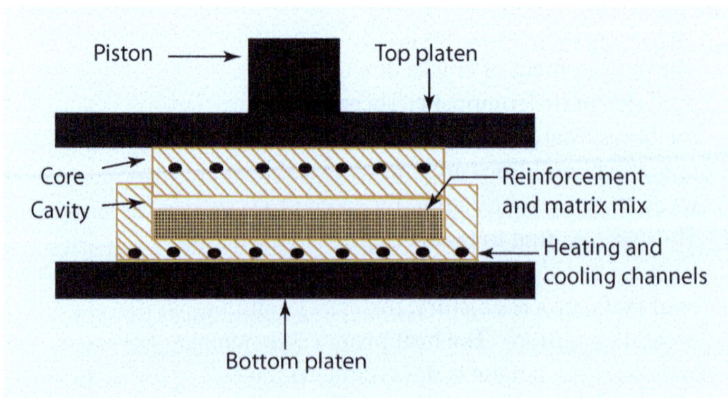

Fig. 8.3 Schematic of compression moulding process

(TMC), and wet layup compression moulding. This is a versatile technology for processing thermosetting as well as thermoplastic composites using any reinforcement in any size and shape. Compression moulding tooling consists of heated metal moulds mounted on a large press.

Moulded materials include those cast of forged steel, iron, and aluminium. Steel moulds are hardened and sometimes chrome plated for enhanced durability. The moulds are heated using steam, hot oil, or electricity. Side cores, provisions for inserts, and other refinements are often employed.

This process is akin to matched die moulding and necessarily uses a hydraulic machine for closing and opening the mould. This process is specifically used for thermoplastic and thermosetting matrices. Short fibres, long fibres, fabrics, or whiskers may be used as reinforcements in compression moulding. The pressure requirement in the compression moulding process is high, and machines are available from 10 ton to 5000 ton hydraulic pressure and more, depending upon the size of the product. The quality of composite products coming out of this process is excellent. Due to the high pressures used, the compaction and volume fractions of reinforcement and matrix are well controlled.

Compression moulding is the universal moulding processes in composites, and it allows better control of the processing parameters. The one drawback of this process is that it is not suitable for very high production rates. Although multi-daylight (the height between the both the platens of hydraulic machine), hydraulic machines are available, the number of daylight is limited to six to ten; therefore, the production rate is limited. Toray Industries Inc. and Toyota Motors, both in Japan, have developed a fuel cell car (shown in ◘ Fig. 8.4) that uses carbon fibre thermoplastic composites that are stake framed and formed in a compression moulding machine.

Compression moulding is also carried out using a solution impregnation method. This involves dissolving the thermoplas-

◘ **Fig. 8.4** Car having carbon fibre-reinforced thermoplastic composite stake frame [11]

tic resin in a solvent. This slurry of thermoplastic and solvent is used as the matrix to impregnate the reinforcement. Later on—during compression moulding—the solvent is evaporated to obtain the moulded composite product. The advantages of this process are the following.

- A very high molecular weight of amorphous thermoplastics can be easily dissolved and processed.
- This process allows the processing of thin films of thermoplastic with micro and/or nano filler.

The disadvantages of this process are the following.

- The complete recovery of solvent is very difficult, making the process costly.
- Use of a solvent makes the process hazardous.
- The complete evaporation of the solvent is difficult.

The disadvantage of compression moulding is that the initial investment cost is very high, which means that normally a large number of mouldings are required to be produced in order to reach a breakeven point and ensure viability of the process. With press-moulded prepregs, there is a limit on the size of mouldings due to the high cost and difficulty manufacturing the large metal moulds and hydraulic presses. In recent years, the development of SMC-type compounds, which are moulded at lower pressures (called LPMC for low-pressure moulding compound), has allowed large parts to be produced economically. However, SMC is essentially a medium- to high-volume process. The production rate is also low as this uses a batch process.

8.1.3 Autoclaving

- *Processing Technology*—Autoclaving is an advanced process for composite fabrication. Although it is a costly process, it yields good composites of high quality for aerospace applications. Mostly, autoclaving is used when byproducts are to be removed during the processing. In autoclaving, isostatic pressure on the outer surface of the moulding, the vacuum inside the moulding, and the temperature can be applied simultaneously, as shown in ◘ Fig. 8.5. This process needs prepregs, which are resin pre-impregnated fabric, so that the volume fractions of the resin and the reinforcement are well controlled. The prepregs are placed in a matched die mould or an open mould, depending upon the composite requirements.

  The autoclaving process requires consumables such as a breather cloth, bleeder cloth, release film, sealing putty, and bagging film to complete the process.

  - A *breather cloth* is used for applying a uniform vacuum all over the moulding.
  - A *bleeder cloth* is used for the absorption of residual byproducts.
  - *Release film* is applied between the bleeder cloth and the moulding to prevent them from sticking together.

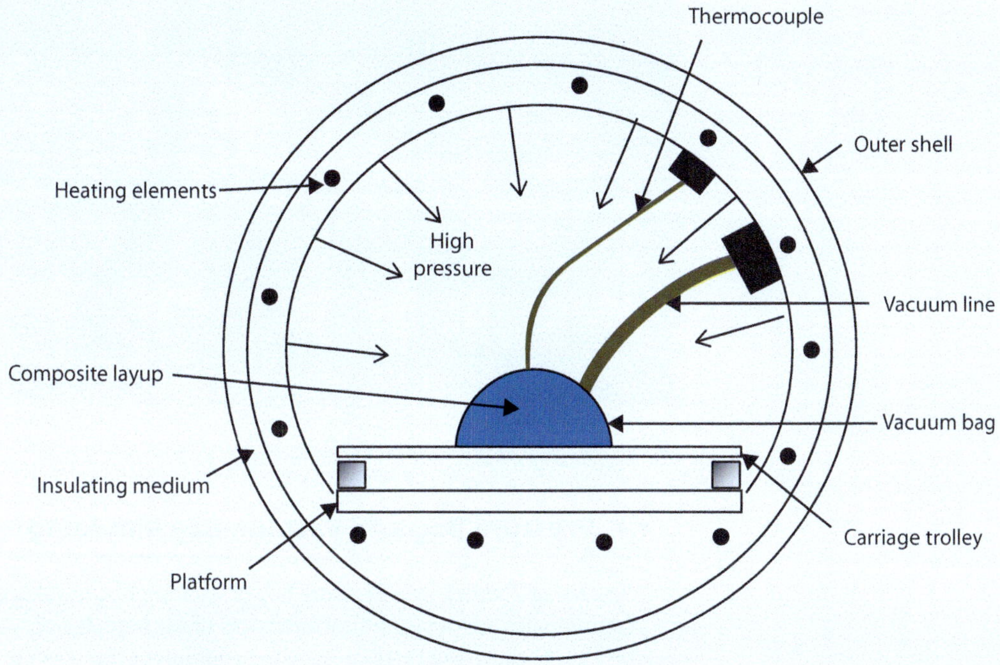

Fig. 8.5 Autoclave process

  - *Sealing putty* is applied to seal the bagging film that covers the entire mould so that the vacuum can be applied.

Autoclaving is a batch process, and the cost of processing is very high due to the cost of consumables involved with each product. It is primarily a process used for thermosetting resin-based composites. The world's largest autoclave (see Fig. 8.6) has been made for the Boeing 787 Dreamliner. The jet has a diameter of 9 metres, length of 30 metres, and gross weight of 920 tons.

- *Processability Parameters*—The processability parameters for autoclaving technology are given here.
  - *Time and temperature*—Time, temperature, and the cure cycle play an important role in autoclaving technology. The temperature at which condensate comes out of the resin and the hold time at this temperature ensure compaction in the composite. Therefore, the processability of the composite depends upon the hold time when the condensate comes out and the cure time for high cross-link density, high glass transition temperature (T_g), etc. Autoclaving is often used for thermosetting composites in which resin releases condensate and byproducts.
  - *Shape and profile of the product*—Sharp edges and undercuts do not favour the processability of composites by autoclaving as they may puncture the vacuum bag.
  - *Form of reinforcement*—Only fabrics and CSM are used as reinforcements in order to allow the processability of the composite in this processing technology.

Fig. 8.6 World's largest autoclave [12]

8

8.1.4 Pressure Bag and Vacuum Bag Moulding

- *Processing Technology*—Pressure bag or vacuum bag moulding technologies are the economic offshoots of the autoclaving process because the costly processing equipment (such as the autoclave) is not used. Instead, a hot air circulation oven is used. Pressure bag or vacuum bag moulding also differs from autoclaving technology because either pressure or a vacuum is applied along with the heat, whereas in the autoclaving process, all three can be applied simultaneously.

  The schematic of the vacuum bag process is shown in Fig. 8.7. In a vacuum bag process, the composite layup or prepreg on a mould is placed in the bag in the same manner as it was placed in the autoclaving process described earlier. Once the bagging is complete, a vacuum is applied, and the mould is placed under the vacuum in the oven for curing the composite.

  In the case of a pressure bag process, the bagging is done in the similar way, but here the pressure is applied inside the product in such a way that the composite layup or prepreg presses against the walls of the mould cavity. The schematic of this process is shown in Fig. 8.8. Later, the mould is placed under pressure in the oven for curing of the composite. The pressure bag technology is used for tubular confined structures in general. This technology is popular for both thermosetting and thermoplastic matrices.

  In vacuum bag technology, a flexible film (PVA, nylon, mylar, or polyethylene) is placed over the wet layup. The wet layup is made by using either thermosetting prepregs or a hand layup process. The layup is placed inside the mould, then the edges at the end of mould and film are sealed, and finally the vacuum is applied. For the vacuum bagging process, a release film is placed over the laminate. This is followed by a bleeder ply of non-woven cloth of nylon, polyester, or other material that absorbs excess resin from the laminate. A breather ply of a non-woven fabric is placed over the bleeder ply to enable a uniform suction of air. After that, a vacuum bag is mounted

Point to Ponder...
Pressure bag technology is used for making carbon or Kevlar fibre-reinforced nylon composite hollow frames for sports bicycles. The damage tolerance of these frames is good.

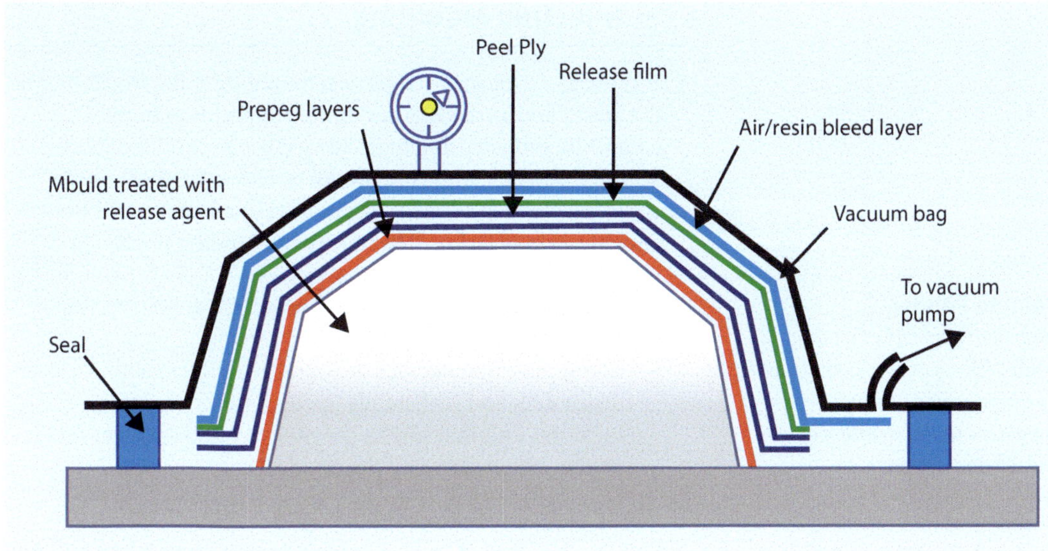

Fig. 8.7 Vacuum bag process [13]

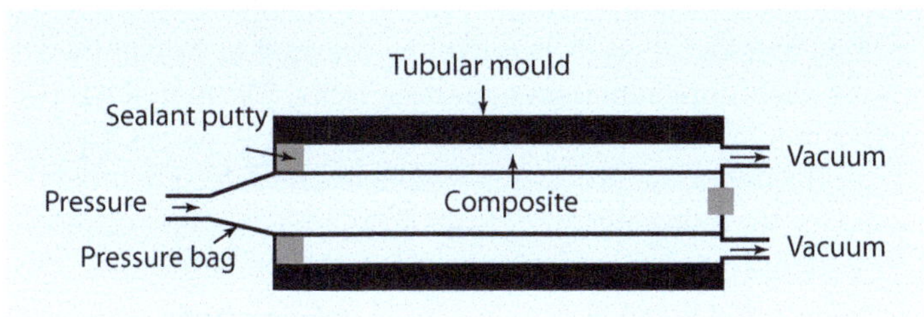

Fig. 8.8 Pressure bag technology

over the entire assembly. Applying the vacuum inside the bag causes an atmospheric pressure that eliminates voids and provides compaction to the laminate.

Vacuum bag processing produces laminates with a uniform quality. The removal of entrapped air inside the laminate reduces the void content. Structures fabricated with traditional hand layup techniques may result in composites that are resin rich, but vacuum bagging eliminates this problem. Additionally, complete fibre wet-out can be accomplished if the process is done correctly. The final results of improved compaction and core bonding are possible with vacuum bag processing.

- *Processability Parameters*—The processability parameters are similar to those explained in vacuum infiltration technology. The difference in vacuum bag technology is that it supports using a high-temperature resin as matrices. This processing technology also supports the processability of thermoplastic prepregs using pressure bag technology.

  The disadvantage of autoclaving and vacuum moulding is the use of excessive consumables that are quite costly and make the product cost also very high. The long time for processing also makes the production slow.

8

8.1.5 Injection Moulding

Point to Ponder...

High heat capacity of a material causes a long time to heat or cool it. Low heat capacity is preferred in mould materials so that it can be heated and cooled fast.

- *Processing Technology*—Injection moulding is the workhorse of the plastic industries but has limited application in the composite industry due to its limited use to produce short fibre-reinforced composite products. However, short fibres could become long staples, depending upon the mould design and injection pressure. The injection moulding process is shown schematically in Fig. 8.9.

  The injection moulding machine has the following important parts:
  - Clamp unit that decides the tonnage of the machine, closes the mould, and withstands the high injection pressure.
  - Injection unit that injects the molten plastic at a very high pressure in order to fill the mould cavity.
  - Heating unit that heats the plastic material before the injection.
  - Control unit that controls all of the processing parameters.

This process uses a specialized mould where the mould deign is quite complex, including runner gates and multiple cavities. While injection moulds are costly because they are used in the continuous process, their durability is an important factor. The mould has two parts: a mould base and mould cavities.

The important processing parameters for injection moulding are the following:
- Temperature from feeder zone to injection nozzle
- Injection pressure
- Hold time
- Locking force
- Mould temperature

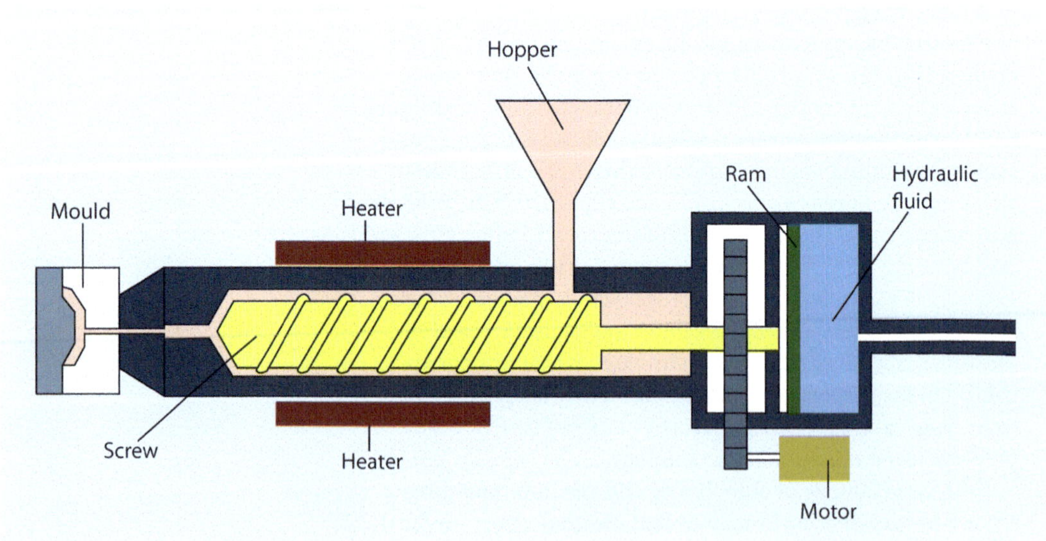

Fig. 8.9 Injection moulding process [14]

- *Processability Parameters*—The following are the processability parameters for processing polymeric composites using a contact layup process.
  - *Type of matrix resin*—Only thermoplastic matrices are processable by injection moulding technology. Matrices that degrade at their melting temperature are not processable by injection moulding.
  - *Melt flow index (MFI)*—The matrix should have high MFI (in excess of 20 g/minute). This enables the filling of the cavity at low pressures.
  - *Form of reinforcement*—Only short fibres, whiskers, and particles can be reinforced in the composite using injection moulding technology. The processability of fabric-reinforced composites is not feasible.
  - *Size and shape of the product*—Products having undercuts and a thick section (>3 mm) are difficult to injection mould.
  - *Injection pressure*—Injection pressures that are lower than the locking pressure are only processable. The injection pressures are calculated based on the projected area of the product.
  - *Design of injection mould*—The high heat capacity of the mould material restricts the rate of production in injection moulding. In order to improve the processability of the composite, cooling the injection moulds is required to be fast.

8.1.6 Pultrusion

- *Processing Technology*—Pultrusion is a manufacturing process for producing continuous lengths of composite with regular and constant cross sections. In this process, a preform of reinforcement is passed through resin and into the moulding die. The moulding die has a heating mechanism to cure the thermosetting resin. The cross section is determined by the opening profile of the die.

  Raw materials used are a liquid resin mixture (containing resin, fillers, and additives) and reinforcing fibres, yarns, rovings, or fabrics. The pultrusion process involves pulling these raw materials through a heated metallic (e.g. steel) die using a pulling device. Therefore, this process becomes continuous. The continuous strand fibreglass roving, mat, cloth, etc. are impregnated in a resin bath, and then it is pulled through a steel die by a powerful tractor mechanism. The heated steel die consolidates the wet reinforcement, sets the shape of the stock, and controls the fibre/resin ratio. Therefore, the profile that comes out of the mould is a hardened and cured composite with regular cross sections. The pultrusion process is used for the bulk manufacturing of beams, bars, I-section and hollow cross-sectional members, etc. A hybrid composite profile is achieved by using hybrid fabric stakes, foams, etc.

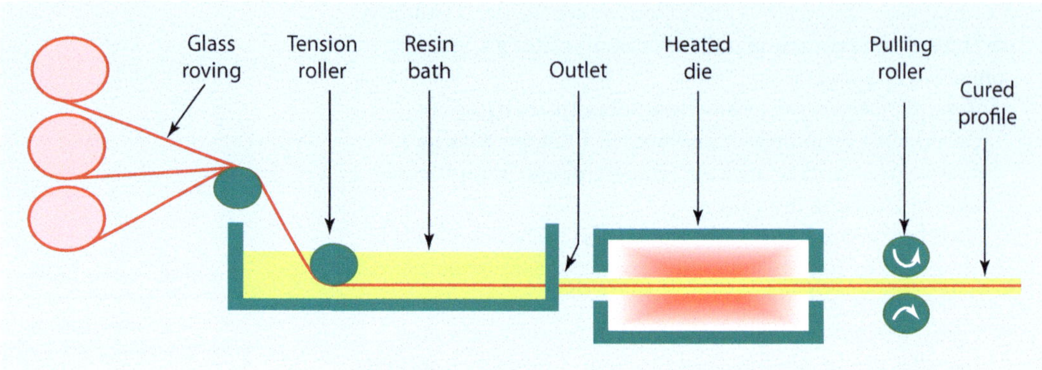

Fig. 8.10 Schematic of pultrusion process [15]

The pultrusion process is a continuous operation that can be readily automated. It is adaptable to both simple and complex cross-sectional shapes. Very high strengths are possible due to the fibre loading. The manufacturing cost in this process is low. The glass rovings, cloth, or mat come out from a wound spool and go through the resin bath through nip rollers, as shown in Fig. 8.10. Then the impregnated glass rovings, cloth, or mat go through the die and adapt to the shape of the profile of the die. At this point, curing also takes place so that the pultruded composite's profile gets cured and the permanent shape, profile, and cross section are formed.

- *Processability Parameters*—The following are the processability parameters for processing polymeric composites using the pultrusion process.
  - *Type of matrix resin*—Room-temperature as well as high-temperature curing thermosetting resins are processable using the pultrusion process. Thermoplastic matrices are not processable by pultrusion technology.
  - *Form of reinforcement*—Only continuous reinforcements (such as fabric, tapes, rovings, tows, and CSM) can be processed in pultrusion technology. Short fibres, whiskers, and particles are not processable by this processing technology.
  - *Mould design*—The mould design needs to support the continuous production; therefore, the composite is processable only when the profile of the cross section is regular, the sliding surface of the mould is smooth, and an adequate heating arrangement for the curing of the product is available.
  - *Residence time in the mould*—The residence time of the green composite in the mould is an important processing parameter. It must be optimum for curing. If the time of residence is less than the required time, the composite would not be cured, and if the residence time is more than required time, the thermal degradation and/or excessive cross-link density may cause a brittle composite.

- *Temperature of the mould*—The temperature of the mould is set in gradients, and processability depends upon the wetting of the reinforcement, followed by the gelling of the resin, and finally by the curing of the composite. The temperature zones must be strictly regulated for optimum properties of the composite.
- *Extraction speed*—The speed of extraction decides the residence time during the pulling process in pultrusion. Therefore, the processability of the pultruded composite depends upon the speed of extraction.

Example 8.2 suggests the processing technology for the fabrication of (a) moulded thermoplastic furniture, (b) hollow seat frame for aircraft, (c) boat hull, and (d) radome for aircrafts.

Answer

(a) *Moulded thermoplastic furniture*—It can be moulded through an injection moulding process.

(b) *Hollow seat frame for aircraft*—The most suitable process would be pressure bag technology.

(c) *Boat hull*— It may be fabricated using vacuum bag technology. A wooden or steel mould may be used for this process. A foam core is also used to increase the stiffness and bring down the density of the boat hull.

8.1.7 Vacuum-Assisted Resin Transfer Moulding (VARTM)

- *Processing Technology*—Vacuum-assisted resin transfer moulding (VARTM) is a closed moulding process for batch production. In this technology, the reinforcement is placed in the cavity of a matched die mould that has injection and outlet port(s). Continuous strand mats and woven reinforcement are laid to dry on the core of the mould. Preformed glass reinforcements are often used for complex mould shapes. The mould is closed and clamped. A vacuum is applied from the outlet port to evacuate air trapped in the reinforcement. A vacuum also is used to enhance the resin flow and reduce void formation. From the injection port, a low-viscosity resin mixed with hardener is injected at a high pressure. In the case of two component systems of resin, metered mixing equipment is used to control the resin/catalyst or hardener ratios that are mixed through a motionless/static mixer and injected into the injection port. The VARTM process is shown schematically at Fig. 8.11. For an optimum surface finish, a gel coat may be applied on the mould surface prior to moulding. Once the mould is completely filled with the resin, the resin will start coming out from the outlet port. Initially, the resin may come out with some air bubbles that had been trapped in the reinforcement. Later, clear resin starts coming out, and at this

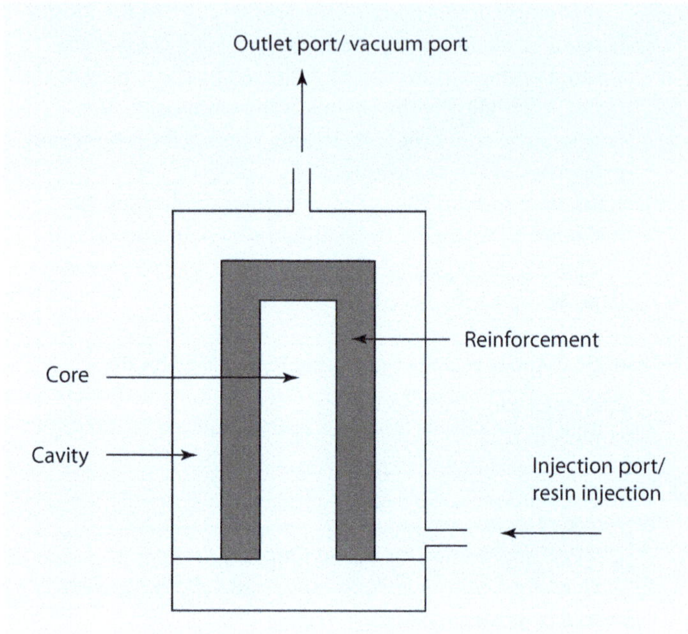

Fig. 8.11 Schematic of VARTM process [16]

juncture, the outlet port is closed first, then the injection port is closed, and finally a high vacuum is applied to evacuate any entrapped air. The mould under the vacuum then is placed inside the oven for curing. After the curing is complete, the finished composite is taken out of the mould.

High-quality parts having a good surface finish on both the sides of the product along with extremely good control over volume fractions of reinforcement and matrix are produced by this method. For example, composite radomes for the aircrafts are made using VARTM process. Also, the composite annulus filler for Rolls-Royce jet engines is made from carbon fibre-reinforced composites using the VARTM process. The VARTM process provides good control over the mechanical properties and surface finishes that are essential for aerodynamics, as shown in Fig. 8.12. VARTM has improved the production rate annulus filler along with a reduction in part weight, resulting in lower fuel consumption and low CO_2 emission.

Advantages of the VARTM process include lower emissions than from open mould processes; high-quality finish; dimensional tolerances within ±100 μm; complex mould shapes; and cabling, duct, pipe, and other fittings that can be incorporated into the product design.

Disadvantages of the RTM process are that high production volumes are not feasible.

- *Processability Parameters*—The following are the processability parameters of VARTM technology.
  - *Type of matrix resin*—High-temperature curing thermosetting resins are processable by VARTM technoogy. Thermoplastic matrices are not processable by this technol-

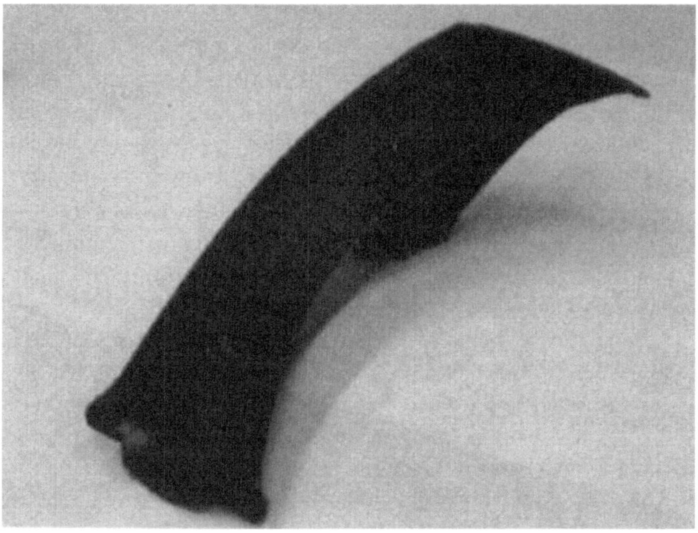

Fig. 8.12 CFRP annulus filler for a Rolls-Royce jet engine

ogy. The processability becomes complex through the resin systems that release condensate or byproducts. The resins that undergo cross-linking by addition polymerization are good to process using VARTM technology.

- *Form of reinforcement*—Only continuous reinforcements (such as fabric, tapes, rovings, tows, and CSM) can be processed in VARTM technology. Short fibres, whiskers, and particles are not processable by this processing technology.
- *Mould preparation*—In order to mould a composite using VARTM, the mould must be cleaned before the layup of the reinforcement and must be airtight after the layup so that while under a high vacuum or pressure, no leak takes place.
- *Application of vacuum and pressure*—The processability is dependent on the evacuation of air from the reinforcement placed in the closed mould so that during the resin transfer no voids or air gaps are present. Also, the resin transfer at a high pressure is an important parameter needed to wet the reinforcement completely.
- *Evacuation of resin*—Resin evacuation is needed for a sufficient time to remove any entrapped air or gases. This must be accomplished so that after curing these gases do not form defects in the composite.
- *Viscosity of the resin*—Viscosity of the resin needs to be low so that it percolates the tightly stacked reinforcement in the mould.

The basic arrangement for VARTM moulding uses a set of matched die moulds and a resin injection machine. The disadvantage of VARTM technologies is that the initial investment cost is much higher, requiring a more expensive mould and a resin injection machine, compared to open moulding. The higher investment cost of VARTM means that the production rate is quite limited.

Typical parts manufactured by VARTM include airframes, fuselages, and automobile parts where tighter tolerances and high mechanized properties are required. The size and shape of the moulding play an important role in the VARTM process, and the cost of big moulds is prohibitive. VARTM is a batch process, and the production volumes are low. There are quicker and more efficient versions of VARTM called RTM Lite, VARI, and Vacflo.

8.2 Processing-Induced Defects in Polymeric Composites

Defects in composites occur during either processing or their service life. Processing-induced defects in composites may occur due to the following reasons.

8.2.1 Unprepared Reinforcement or Matrix

It was discussed previously (► Chap. 4) that the surface of the reinforcement must be hydrophobic for its compatibility with a hydrophobic matrix. If the surface treatment for compatibility with the matrix is poor, the composite will be of poor quality, having low strength, low modulus, and low interlaminar shear strength (ILSS) . Also, the surface treatments are responsible for efficient handling of the reinforcement. If the surface coating of carbon fibre is defective, then, during the filament winding of the carbon fibre, a fuzzing of carbon fibre is observed at the payout eye. Usually, a coating of polyimide or epoxy resin is placed on the carbon fibre surface to make it compatible with the organic resins and easy to handle.

As far as matrices are concerned, all of the components of the matrices (in two- or three-component systems) must be mixed in correct proportion to provide optimum properties. In the cases where a hardner is mixed more, the cross-link density increases, and high exothermic reactions occur, causing a poor quality of composites. Similarly, low hardner contents lead to under or no curing of the composite, affecting its glass transition temperature and mechanical properties. In one component matrix system, such as phenol formaldehyde resin (popularly known as phenolic resin), lumps of resin due to ageing are observed, which cause a poor composite quality. The processing-induced defects for open and closed mould processing are given here.

8.2.2 Improper Storage Conditions

Storage of a resin is important because resin matures with heat and time. The maturity of resin is nothing but the initiation of a low degree of cross-linking, which reduces the flowability of the resin matrix in the composite. Resin matrices are susceptible to moisture absorption. The ingress of moisture causes bulk and surface defects during high-temperature curing due to the formation of water vapours. The accumulation of dust and foreign particles on the surface of the reinforcement or inside the matrix reduces all of the properties of composites.

8.2.3 Wrong Placement of Fibres

Tailoring of the in-plane stiffness of composite laminates can improve the buckling and post-buckling characteristics by redistributing the applied loads. The fibre placement technique used is the core technology underpinning variable stiffness composites. The mismatch between designed and real fibre paths increases as the width of a single tow and the number of tows laid at the same time increase. This causes nonuniformity of the fibre orientation, and some defects such as tow gap and overlap and irregularity of the thickness distribution will occur in composites.

8.2.4 Improper Volume Fractions

Improper volume fractions lead to either resin-rich areas or resin-starved areas, and both of these situations are detrimental to the performance of composites. For the best performance of the polymeric composites, equal fractions of volumes of reinforcement and matrix must be kept.

8.2.5 Uncontrolled Temperature, Pressure, and Time

One of the major challenges in the manufacturing of composites is the efficient handling of the fibres and the application of uniform pressure during the curing of the resin. Fibres are stiff in nature and do not flow. For this reason, continuous fibres are either stacked in layers that are successively impregnated with the resin or formed into a preform, which is placed in a mould where resin is supplied under pressure to wet the fibres. In addition, getting a high-volume fraction of fibres in the composites is a major concern. Properties of composites depend upon the appropriate curing of the matrix. The pressure, temperature, component geometry, and distribution of pressure over the surface of the component are some of the deciding factors for maintaining the quality and reliability of the manufactured component.

The temperature gradient (T_g) through the thickness of composites occurs during their curing process. The insulating characteristic of composites restricts the heat transfer and increased level of residual stresses from forming. However, this increases the risk of fibre waviness and wrinkles, leading to a risk of defects in the composite. Thick laminates are subjected to combined effects of cure exotherm, leading to high stresses in off-axis plies, which increases the risk of waviness and wrinkles in the plies (especially with complex shapes).

8.2.6 Improper Closing of the Mould

Improper closing of mould can lead to dimensional instability of the cured product, and also it may lead to resin oozing out upon curing. The oozing out of the resin leads to an imbalance in the volume fractions of the reinforcement and the matrix. Thus, the composite will have low mechanical properties.

8.2.7 Demoulding Temperature

A moulding (product) must be taken out of the mould at room temperature or at a prescribed demoulding temperature. If the moulding is removed from the mould at a higher temperature than the demoulding temperature, warpages may occur in the product due to the generation of residual stresses. The residual stresses are caused due to a thermal nonequilibrium between the reinforcement and the matrix at high temperatures, even though curing is complete. The reason for thermal nonequilibrium is the difference in linear thermal expansion between the reinforcement and the matrix. This problem is observed most in carbon fibre-reinforced composites because the conductivity of carbon is much higher than the conductivity of the matrix. Therefore, the coefficient of linear thermal expansion of the carbon fibre is more than the matrix.

8.2.8 Post-fabrication Finishing Operations

In many composite products, there is a need to carry out post-fabrication finishing operations, such as surface painting, metalizing, and fitment of assembly. All of these operations demand care in handling so that no damage to the surface is caused. Many a time, surface defects are not visible, but they remain in the form of microcracks or surface crazing. Microcracks or surface crazing leads to debonding and delamination, which eventually causes failure of the composite.

8.3 Future Composite Manufacturing Technologies

Composites are primarily used for their good properties and to replace conventional materials. In order to make composites competitive against metals (especially steels), cost reduction has to occur in three areas: *nonrecurring costs*, *recurring costs*, and *direct operating costs (DOC)* (e.g. durability, maintainability, reliability, and reparability). The DOC are a major issue because many high-end products (aircrafts, automobiles, yachts, etc.) depend heavily on the use of composites. There is always a debate on the cost-effectiveness of composites for commercial use; however, such debates can only be put to an end with the development of newer cost-effective processing technologies. The cost-effective processing alone will not be able to bring down the manufacturing costs until nonrecurring costs and DOC are also brought down. Thus, a life-cycle cost analysis should be conducted along with the traditional trade-off between the overall properties of composites and their cost. Current developments in the area of advanced moulding processes of composites are given here.

8.3.1 Stitched Preform Followed by VARTM

This process involves the *stitching* of reinforcement fabric on the mould for snug fitting followed by VARTM to fabricate the composite part. Small- to medium-sized stitched preform-based composite products are in use in composite industries. This method is not cost-effective for small- to medium-sized thin parts; to take full advantage of this method, the parts must be thick and large. For the acceptable use of this technology for big products, an appropriate automatic stitching machine is needed because the preforms are of different shapes and thicknesses. The stitching of preforms of various skin thicknesses—ranging from less than ¼ inch to more than 1 inch—is required.

8.3.2 Continuous Sandwich Panel

Continuous sandwich panels may be obtained in thermoplastic and thermosetting composites. The thermosetting composite panels may be produced through a pultrusion process, whereas thermoplastic sandwich panels may be manufactured through an extrusion process. However, these processes are limited to making flat, constant sandwich panels. Floor panels, door panels, partitions, etc. are the major products from flat sandwich panels. Continuous sandwich panels of complex shapes and variable thickness are produced in abundance for such applications.

8.3.3 3-D Weaving

3-D *weaving* of reinforcing yarn is a machine-dependent operation. The cost of 3-D weaving is prohibitively high due to the weaving machine cost. The composite product made from 3-D preform is similar to an isotrophic material having good mechanical properties in all three directions.

8.3.4 Mechatronics

Composite parts for high-end applications (aircrafts, missiles, etc.) in general are made in batches, so production rates are usually low. *Mechatronics* is a semiautomated process for the design and fabrication of any product. As a result, this process may be used for the manufacturing of composite products for high-end application. The mechatronics process ensures high-quality products.

8.3.5 Co-curing Technology

In *co-curing technology*, two dissimilar matrix composites are cured together in a single assembly. This process forms the completely assembled product in a single operation, reducing the costs of assembly and time. However, there can be problems with the compatibility of subassemblies during curing. Also, the formation of induced or residual stress must be avoided to ensure the quality of the product. Co-curing is carried out by controlling the curing parameters, such temperature, time, and pressure.

To Be Remembered

1. Closed moulding processes are those in which all or part of the product is processed in a closed vessel or chamber.
2. Matched die moulding has a matching core and cavity.
3. Compression moulding technology is the workhorse of the closed moulding process in composites.
4. Autoclaving and vacuum bag processes are used where condensate and byproducts come out during the curing process.
5. Costs of consumables increase the processing cost in autoclaving and vacuum bag technologies.
6. Injection moulding technology is the most popular manufacturing technology for the production of short fibre-reinforced thermoplastics..
7. Important parts of injection moulding machines are the clamping unit, injection unit, heating unit, and control unit.
8. Pultrusion is the fastest manufacturing technology for thermosetting resin-based composites.
9. Vacuum-assisted resin transfer moulding (VARTM) is used for products that require good control over volume fractions of the reinforcement and the matrix. VATRM alsp provides a good surface finish on both sides of the composite product.

Exercises

1. How many layers of glass fabric are required for making a planar laminate of 3 mm thickness with ideal volume fractions? The thickness and arial density of glass fabric are 0.33 mm and 425 GSM (g/m^2), respectively. Densities of glass and epoxy resin are 2.54 g/cm^3 and 1.2 g/cm^3, respectively.
2. Write down the processing parameters for the injection moulding process.
3. Why is the temperature of an injection nozzle kept at the maximum temperature compared to other zones of an injection barrel?
4. A 10-mm-thick ballistic panel has been fabricated out of Kevlar-reinforced polyetherimide composite. Estimate the numbers of

layers of Kevlar fabric in this panel. The density of Kevlar fibre, nominal thickness of Kevlar fabric, density of polyetherimide, and area density of Kevlar fabric are 1.44 g/cm^3, 0.40 mm, 1.27 g/cm^3, and 300 g/m^2, respectively.

5. Find out the density of the matrix in a 6-mm-thick carbon fibre-reinforced composite. The density of carbon fibre, nominal thickness of carbon fabric, and area density of carbon fabric are 1.8 g/cm^3, 0.25 mm, 1.27 g/cm^3, and 225 g/m^2, respectively.
6. Why is compression moulding suitable for the processing of thermoplastic as well as thermosetting composites?

References

1. Murphy, J., *The Reinforced Plastics Handbook*, Elsevier Advanced Technology, Oxford, 1994.
2. Monk, J.F., *Thermosetting Plastics: Practical Moulding Technology*, George Godwin, London, 1981.
3. Kia, Hamid G., *Sheet Moulding Compound: Science and Technology*, Hanser, 1993.
4. CIRRIA, *The Use of Epoxy, Polyester and Similar Reactive Polymers in Construction*, Vols. 1, 2, and 3, USA, 1993.
5. Sigg, A., De Oliveira, R., Hochman, L., Cazes, T., and Michaud, V., *Process induced defects in thick composite laminates*, Laboratoire de Technologie des Composites et Polymères (LTC), Ecole Polytechnique Fédérale de Lausanne (EPFL). MXH143, Station 12, CH 1015 Lausanne, Switzerland.
6. Gürdal, Z.and Olmedo, R., *Composite laminates with spatially varying fiber orientations: variable stiffness panel concept*. Proceedings of the AIAA/ASME/ASCE/AHS/ASC 33rd structures, Structural Dynamics and Materials Conference, Dallas, TX, Vol. 2, AIAA 92-2472, 1992, p. 798.
7. "Reinforced Thermoplastic Composites for Washing Machine Tubs," *FRP Today*, 14 (3), March 2014, p. 18.
8. Gürdal, Z. and Olmedo, R., "In-plane response of laminates with spatially varying fiber orientations: variable stiffness concept," *AIAA Journal*, Vol. 31, No. 4 (1993), p. 751.
9. Olmedo, R. and Gürdal, Z., *Buckling response of laminates with spatially varying fiber orientations*, Proceedings of the AIAA/ASME/ASCE/AHS/ASC 34th Structures, Structural Dynamics and Materials Conference, La Jola, CA, Vol. 1, AIAA 93-1567, 1993, p. 2261.
10. Holloway, J., *Siemens unveils world's largest wind turbine blades*, http://www.gizmag.com/worlds-largest-wind-turbine-blades/23578/, *2012.*
11. *"Carbon Fibre-Reinforced Thermoplastic used in Automobile Structural Parts for the First Time in the World," FRP Today, 14 (12), December 2014, p. 34.*
12. *"World's Largest Autoclave Completed for Boeing 787," FRP Today, 14 (12), December 2014, p. 19.*
13. Rajpurohit, A., "Fiber Reinforced Composites: Advances in Manufacturing Techniques," *FRP Today*, 14 (04), April 2014, p. 22.
14. *Stages of injection moulding and design technology*, http://www.design-technology.org/injectionmoulding2.htm.
15. *Pultrusion*, Wikipedia.
16. *"Development of Composite Annulus Filler for Rolls Royce Jet Engines," FRP Today, 14 (3), March 2014, p. 19.*

Characterization and Testing of Polymeric Composites

A. K. Ghosh, M. Dwivedi, *Processability of Polymeric Composites*,
https://doi.org/10.1007/978-81-322-3933-8_9

The complexity of damages in composites is closely related to anisotropy and includes various mechanisms, such as interfacial debonding, matrix cracking, ply cracking, fibre breakage, and so on. It is critical to understand the evaluation of damage accurately in order to assess and predict the life of polymeric composites. From the discussion, in the preceding sections, it is understood that the composite consists of laminae, and the laminae consist of fibres and a matrix. In order to understand and assess the behaviour of polymeric composites, it is important to understand the performance of these materials under different loading conditions and operational environments. The performance of polymeric composites is dependent on factors like loading conditions (such as cyclic or static), temperature, humidity, volume fractions, exposure to radiation, and the chemical environment. The assessment of the mechanical behaviour may be carried out based on the loading conditions and service environment [1, 2].

The characterization and testing of any material are used to establish a complete understanding of it. Inherent defects in materials and process-driven discrepancies are common in polymeric composites. The characterization and testing help in identifying any flaw in the composites and, in turn, ensure their quality. It is also important to characterize and test polymeric composites to generate the experimental data used to design the product and also to predict its performance.

The characterization and testing methods must be easy to apply and provide accurate results. New instrumentation techniques and automation have made the results of characterization and testing very accurate and easy to interpret. This had led to an easier correlation of the results. Because the characterization and testing equipment produced today are microprocessor controlled, they are very versatile and accurate and have excellent capabilities. This chapter delves in detail on the characterization and testing of composites.

Example 9.1 Cite two examples of processing-driven defects in polymeric composites.

Answer

In a hand layup process, the orientation of reinforcement cannot be repetitive (product-to-product), and this leads to some variations in the properties obtained. Similarly, the composites processed using autoclave moulding technology, ineffective suctionning or vacuuming, or oversaturation of the bleeder cloth lead to defective composite parts.

9.1 Sample Preparation

Sample preparation is the most important and critical activity because it will enable decisions to be made on the acceptance and qualifications of the product, design values, and process design. Sample preparation is not limited to its physical dimension of standard. It is also important in terms of raw materials, processing technologies, and processing parameters. There are two types of approaches for sample preparation: bottom-up and top-down.

1. *Bottom-Up Approach*—In this approach, the sample is moulded or fabricated directly into its final shape or configuration from raw materials. For example, injection-moulded thermoplastic composite samples are moulded in a matched die mould using compression moulding technology or VARTM. These samples have very high dimensional accuracies.
2. *Top-Down Approach*—In this approach, a composite test laminate is fabricated using any of the processes mentioned in ▶ Chaps. 5 and 6. These test laminates are further machined to the dimension of the test samples per their standards. These machining operations may vary from manual cutting by hacksaw to milling, turning, honing, grinding, etc. The secondary machining operations expose the machined surface of the composite to the environment. These secondary operations—more often than not—are manual operations that sometimes cause defects in the composite test sample. For making samples for SEM and TEM where the morphology of a fractured or desired bulk surface need to be studied, this is the only suitable approach.

9.2 Characterisation

The characterization of any material is established based on its morphology or molecular structure. It's charcterization is an intrinsic property of the material. In the case of polymers, character is established based on the properties of polymeric molecules, interphase, and physical form of reinforcement.

9.2.1 Thermal Characterization

Thermal analysis has a unique role in the characterization of polymeric composites. This is effectively used for measuring various morphological and structural transitions in a polymeric composite with respect to temperature. The most important structural transitions are the glass transition temperature, crystallization, melting temperature, heat absorption, evolution, and degradation. Thermal characterization may be carried out rapidly and reveals the very basic nature of the polymeric composite [3–5].

1. *Differential Scanning Calorimetry (DSC)*—DSC provides both quantitative and qualitative data on the material. This technique may be used under a wide variety of conditions, such as different heating rates, different sample sizes, testing medium (air or inert gas), and sensitivity. This is carried out as per ASTM D3418 at a selected heating rate. In DSC, the sample and reference materials are heated to the same temperature, and their temperature is kept the same at any time. The peaks in a DSC thermogram are generated due to thermodynamic changes, releasing or absorbing heat measured in mcal. The

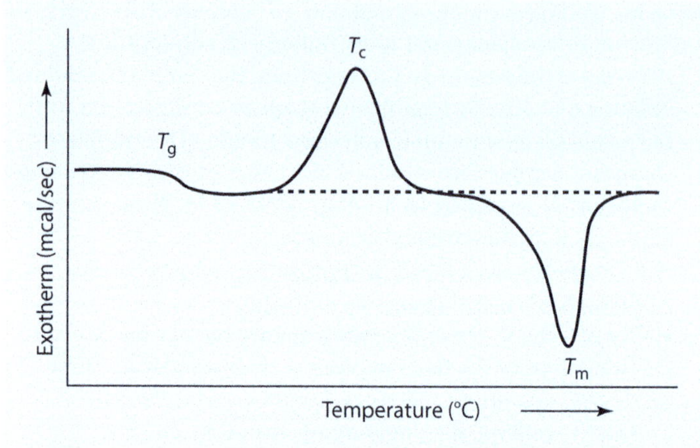

Fig. 9.1 Typical DSC thermogram of an amorphous thermoplastic composite

9

absorbed or released heat is recorded in the form of peaks with reference to temperature in the DSC thermogram. The samples are usually prepared in the weight range of 5–7 mg. The test may be carried out from −170 °C to as high as 800 °C. The results are obtained in the form of heat flow (mcal/sec) and temperature for transitions. In DSC, the area under the peak is directly proportional to the heat involved in the peak formation; therefore, it can be calculated from the thermogram. A typical thermogram of a DSC of thermoplastic composite is given in Fig. 9.1

Example 9.2 Why in semicrystalline thermoplastics does cold crystallization occur after it attains the glass transition temperature?

Answer

The polymeric chains are relaxed at the glass transition temperature due to heat energy. This occurs due to a loosening of entangled polymeric chains in the polymer. These polymeric chains will try to rearrange themselves in a more stable configuration. The arranged structure of polymeric chains causes crystallization. That's how a small degree of crystallinity in semicrystalline thermoplastics occurs.

2. *Differential Thermal Analysis (DTA)*—The results of DTA are obtained from a change in temperature (ΔT) and the temperature of transition. In DTA, the sample and reference materials are heated to the same temperature, and their temperatures are kept equal at any time. A typical thermogram of the DTA of a thermoplastic composite is given in Fig. 9.2

   The sample absorbs or releases heat due to exothermic or endothermic reactions. Accordingly, the sample will absorb heat from the heating source. The difference in temperature supplied to the sample with respect to the reference material is

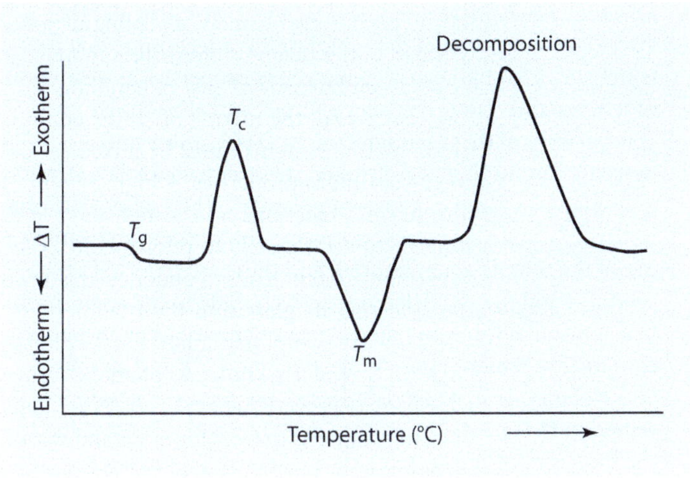

Fig. 9.2 Typical DTA thermogram of semicrystalline thermoplastic composite

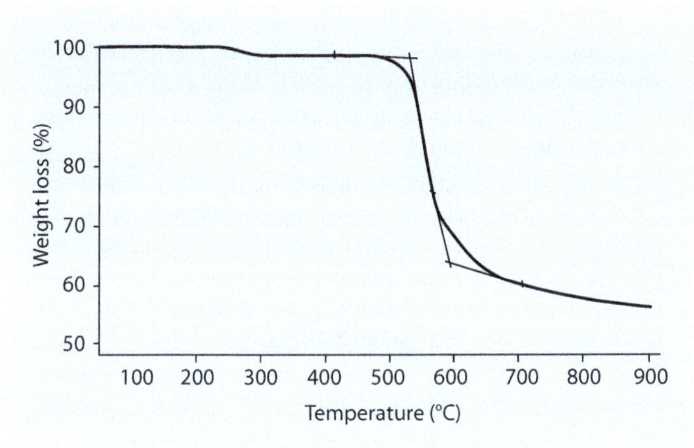

Fig. 9.3 Typical TGA curve of high-temperature thermoplastic composite

recorded in the form of peaks caused due to the temperature difference. The temperature difference is recorded in the form of peaks with reference to the temperature in the DTA thermogram. The samples are usually prepared in the weight range of 5–7 mg. The test may be carried out from −170 °C to as high as 2400 °C [3, 6, 7].

3. *Thermogravimetric Analysis (TGA)*—TGA is carried out to learn the response of the sample in terms of weight loss with respect to temperature. The weight loss occurs due to thermal degradation. The weight loss is measured under an inert or oxidative environment as needed. TGA is carried out as per ASTM D3418. The samples are prepared in the weight range of 5–7 mg. The test is carried out up to 850 °C. The results are plotted as a curve of weight loss with respect to temperature. A typical TGA curve of a high-temperature thermoplastic composite is given in Fig. 9.3.

4. *Dynamic Mechanical Analysis (DMA)*—DMA provides the viscoelastic response of the samples. It is a technique that is widely used to characterize a material's properties as a function of temperature, time, frequency, stress, and atmosphere or a combination of these parameters. In DMA, a sinusoidal deformation through a controlled stress or a controlled strain is applied to a sample of known geometry. The sample deforms under this controlled stress or strain. The stiffness and damping of the sample are measured, and these are reported as the modulus and the tan delta, respectively. Due to the application of a sinusoidal force, the modulus is expressed as an in-phase component, which is the *storage modulus* (*G*'), and an out-of-phase component, which ismthe *loss modulus* (*G*"), as represented in ◘ Fig. 9.4. The storage modulus and loss modulus are the measures of the sample's elastic behaviour and lossy behaviour, respectively. The ratio of the loss modulus to the storage modulus is the tan delta, and this is often called *damping*, which is a measure of the energy dissipation of a material.

   In DMA, an accurate glass transition temperature (T_g) can be measured. The DMA curve of an E-glass fabric-reinforced polyetherimide is shown in ◘ Fig. 9.5. Here, a large drop is seen in the storage modulus at 219.82 °C. A concurrent peak (in dotted lines) of tan delta is also seen.
5. *Degree of Cure*—The transformation of uncured or partially cured fibre-reinforced thermosetting polymers (such as prepegs) into a hardened form is called curing. Curing takes place at specified temperatures and pressures for a predetermined length of time. In curing, a chemical reaction takes place, and the polymer matrix is cross-linked through covalent bonds. The cross-linking is an irreversible reaction. The *cure temperature* initiates and sustains the chemical reaction that

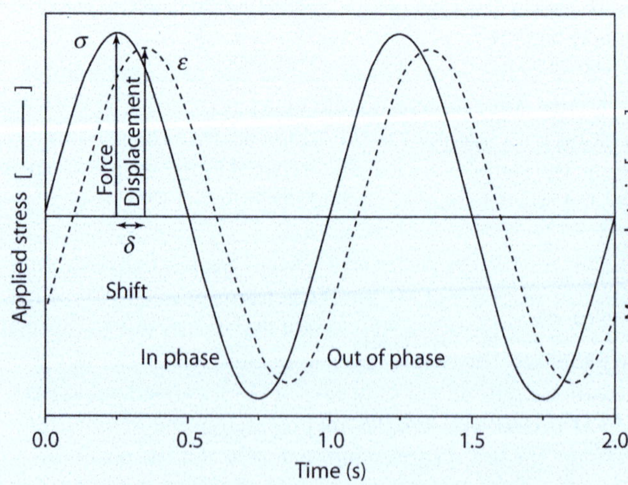

◘ **Fig. 9.4** Typical TGA curve of a high-temperature thermoplastic composite

9

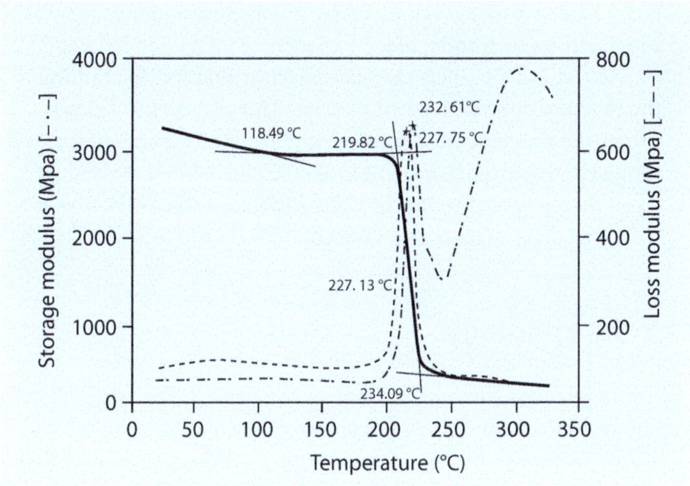

Fig. 9.5 Typical TGA curve of a high-temperature thermoplastic composite

transforms the uncured or partially cured resins into a fully cured solid resin. The *cure pressure* provides the necessary force that is required for the flow and percolation of the viscous resin around fibres. The cure pressure also provides a desired shape and compaction to the composite. The amount of temperature and pressure as well as their duration significantly affects the performance of the composites. The length of time, with respect to temperature and pressure, required to cure composites is called the *cure cycle*. The cure cycle is dependent on the resin chemistry, morphology, thermal activity, and additives (such as inhibitors or accelerators) [8].

The curing cycle has various parameters. For room temperature curing resin systems, an important parameter is gel time. It is tested for initiation of curing. It is a test which determines the time duration for on set of cross-linking, i.e. gelation of resin system. Time duration of gelation varies with temperature. Gelation time reduces with increase in temperature. Gelation leads to cross-linking of room and a solid mass is formed.

For a high-temperature curing resin system, the process of gelation leading to cross-linking begins at the same temperature from which the chemical reaction for cross-linking is initiated. Once the cross-linking is completed, the resin will twin into a solved mass. The cross-linking reaction is an irreversible reaction.

Full cross-linking causes complete curing of the resin. It can be said that a full degree of cure signifies a complete cross-linking of the matrix resin.

6. *Thermal Conductivity*—The thermal conductivity of polymeric composites is quite poor—except for carbon fibre-reinforced composites. Even in carbon fibre-reinforced composites, the heat conductivity is limited due to presence of a nonconducting resin matrix. The low thermal conduc-

tivity of composites is caused by a low atomic density, weak interactions or bonds, etc.

ASTM D5930–09 is the standard test method for finding the thermal conductivity of plastics. This test method determines the thermal conductivity of plastics or composites over a temperature range from −40 to 400 °C. The thermal conductivity of thermoplastics, thermosets, rubbers, and composites is in the range from 0.08 to 2.0 W/m.K.

9.2.2 Morphology

Morphology is the study of the shape, size, texture, and phase distribution of the constituents in the composite material.

1. *Visual Evaluation*—The morphology of composites is revealed by visual inspection. A patchy appearance on the composite implies that the volume fractions of resin and reinforcement are not even. Moreover, some stain marks indicate hot spots where the resin became over cured, leading to degradation. Such areas are mechanically weaker than the rest of the composite.
2. *Scanning Electron Microscopy (SEM)*—SEM is used to analyse the topography of any material (e.g. composite sample). Its depth of penetration is small (approximately 30 nm). SEM can achieve a resolution better than 1 nm. The surface of the composite needs to metallized by coating it by gold. The scanning electron microscope uses a focused beam of high-energy electrons to generate a variety of signals at the surface of the solid specimens. A typical SEM micrograph of a carbon fibre is shown in Fig. 9.6. The signals that derive the electron to gold-coated composite interactions reveal information about the sample, including the external morphology (texture), composition, crystalline structure, and orientation of the materials making up the sample.

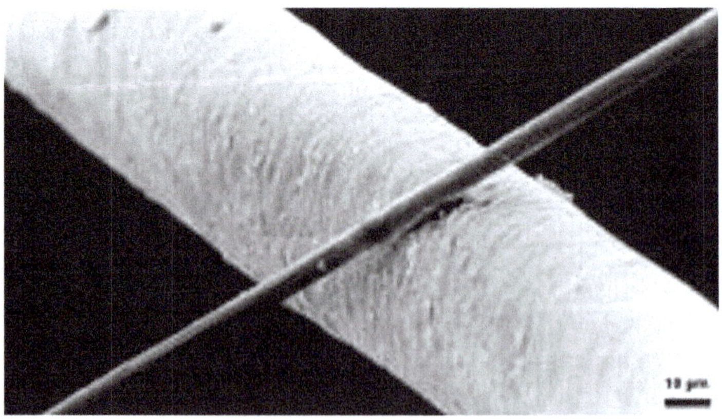

Fig. 9.6 Typical SEM micrograph of a carbon fibre (black in colour) in comparison to a human hair [9]

3. *Transmission Electron Microscopy (TEM)*—In TEM, a beam of electrons is transmitted through an ultra-thin specimen so that an image is formed from the interaction of the electrons transmitted through the specimen. The sample preparation is a specialized activity in TEM. In a sample preparation, a *microtome* is used to cut the sample and coat its surface with carbon. It helps the electron beam to travel and scan below the surface of the sample without creating a reflection outside the surface. The depth of penetration is high in TEM, and it provides information on morphology even below the surface of the composite. TEM provides finer details of the morphology—up to 1 nm. A typical TEM micrograph of a nanoclay-reinforced composite is shown in ◻ Fig. 9.7.
4. *Photometer and X-Ray Diffraction Technique*—In a *photometer technique*, a thin composite laminate is place in front of monochromatic light having a uniform intensity. This technique is applicable for thin composites only. The intensity of transmitted light is measured by a photometer. The optical transmittance is reproducible for uniform composites having similar constituents and volume fractions. The intensity of transmittance is the function of V_f. For a thick section of composites, an *X-ray diffractometer method* is employed. In ths technique, a standard X-ray generator and a standard proportional counter are used. The distribution of the relative intensity of an X-ray diffraction pattern is obtained and recorded through a pulse height discriminator and a rate metre.

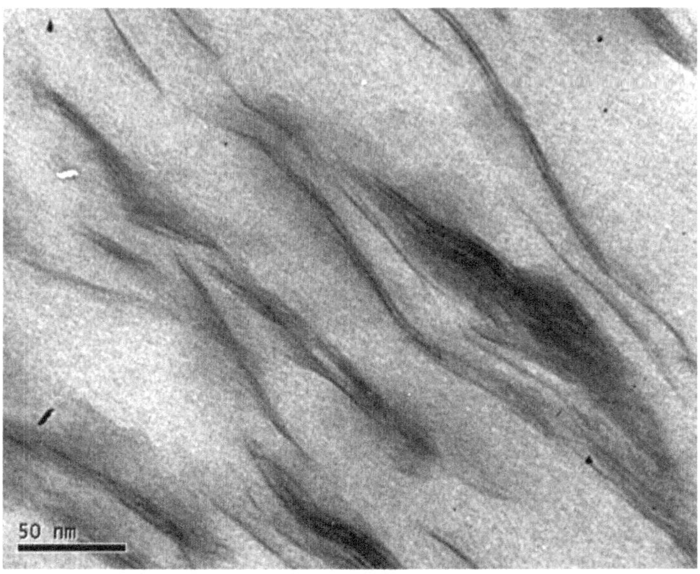

◻ **Fig. 9.7** Typical TEM micrograph of a nanoclay-reinforced composite [10]

9

9.3 Testing

Testing of any material reveals its intrinsic property at a macro level. Polymeric composites are gaining significance; therefore, it is important to test their properties for the design and development of suitable products.

9.3.1 Physical Testing

1. *Coin Tapping Test—Coin tapping* is a qualitative test used to check the compactness of the composite. It is used as the first measure to check the quality of the composite. This test detects only large-sized delamination or under curing of the composite part. A fully cured and good-quality composite will give a sharp sound by tapping the edge of a coin on it. Carbon fibre-reinforced composites give a metallic sound upon coin tapping. If composite has major delaminations and defects, a thud will be heard upon coin tapping.
2. *Measurement of Density*—The specific gravity is measured by dipping samples in water using the ASTM D792A method. In this method, samples are prepared in the form of small pieces. The weight of each sample is taken in air and in water using cotton thread. Precaution is taken for attachment of any air bubbles when the dipping the sample in distilled water. For weighing, a manual chemical balance is used. The weight of the sample is taken by dipping the sample into distilled water by suspending it with the help of a thin cotton thread. The distilled water is filled up to 2/3rds of the volume of a 100 ml beaker. A loop is made on a cotton thread so that by using the same thread and loop all samples can be dipped into the beaker and their weight can be taken. The specific gravity is calculated using Eq. 9.1

$$\text{Specific Gravity} = \frac{W_{\text{Air}}}{W_{\text{Air}} - W_{\text{Water}}} \tag{9.1}$$

where:
- W_{Air} = Weight of laminate (in g) in air
- W_{Water} = Weight of laminate (in g) in water

3. *Measurement of Void Content*—Voids are formed in the composite as a result of gases and volatiles that evolved during processing and are trapped in the matrix. This is undesirable and has has an adverse effect on the mechanical properties of composites. As the void content increases, an increase in the scatter of the strength data is observed. A strong correlation between large voids (area $\geq$0.03 mm^2) and a detrimental effect on the mechanical properties is found. Voids assist in the crack propagation in resin-rich interplay regions. Increased void content increases a composite's susceptibility to water penetration and weathering. Voids in composites are measured using ASTM D 1505 and ASTM D 3800.

The void content of a composite significantly affects its physical, mechanical, thermal, and electrical properties. The measurement of void content is carried out using ASTM D2734. The volume fraction of void content (V_v) is measured by using Eq. 9.2.

$$V_V = \left(\rho_{ct} - \rho_{ce}\right) / \rho_{ct} \tag{9.2}$$

where:

- V_v = Volume fraction of voids
- ρ_{ct} = Theoretical density of composite
- ρ_{ce} = Experimental density of composite

4. *Ultrasonic Through-Transmission Test*—In this test, pulsed waves representing the "sound" are transmitted through the sample. In *through-transmission* (or attenuation) mode, a transmitter sends an ultrasound through one surface, and a separate receiver detects the amount that has reached it on another surface after travelling through the medium. Imperfections or other conditions in the space between the transmitter and receiver reduce the amount of sound transmitted, revealing their presence. A diagnostic machine displays these results in the form of a signal with the amplitude representing the intensity of the transmission and the distance representing the arrival time of the transmission. This test is carried out using an ultrasonic flaw detection tester as per the ASTM E2580 standard. The energy of the test varies from 10 dB to 30 dB based on the thickness of the sample with dry contact and rubber probes at a frequency of 1.0 MHz (approximately).
5. *Evaluation of Constituent Content*—The stiffness and strength properties of a composite are strongly dependent on the fibre volume fraction (V_f), and this parameter constitutes an important quality measure. For low-fibre volume fractions, the damage is in the form of matrix cracking and interface debonding, while delamination dominates concerns for specimens with higher fibre volume fractions. The strength parameters vary in direct proportion to the fibre volume fraction in composites.

   The dynamic properties of composites also are affected by fibre volume fraction. The storage moduli of unidirectional composites increase as the fibre volume fraction increases. The fibre volume fraction of a composite is determined using a chemical matrix digestion method as described in ASTM D3171, the burn test as described in ASTM standard D2854, or photo micrographic techniques. In the *burn test* method, small pieces of glass fibre-reinforced composite of known weight are kept in crucibles and heated in the furnace at 650 °C for 4–6 hrs. The crucibles are then cooled to room temperature, and the residual glass fibre is weighed. For glass fibre composites, the burn test is used because glass fibres are resistant to oxidation at the temperature required for burning off the matrix (500–600 °C).

In carbon epoxy composites, carbon fibre content is calculated using a *chemical decomposition* method. In this method, a carbon fibre composite is placed in concentrated nitric acid, and the mixture is boiled. The matrix gets dissolved in concentrated nitric acid, and carbon is decanted as a residue, washed, and dried. The dried carbon fibre is weighed to calculate V_f. In the case of carbon fibre-reinforced polyimide composites, hydrazine hydrate is used in place of concentrated nitric acid. In the case of thermoplastic composites, solution methods are used. In PP matrix composites, the PP is dissolved in xylene.

There is also a nondestructive testing (NDT) method for evaluating the reinforcement content in the composite. This technique uses selective radiation where samples can be tested for glass content in 2 minutes with an accuracy of ±1%. The selective radiation can separate out glass fibre from all other fillers, additives, and pigments used in composites.

9.3.2 Mechanical Testing

Mechanical testing (tensile test, flexural test, and impact test, etc.) of composites refers to the determination of material properties through tests conducted on suitably designed specimens. The understanding of material response over an entire range of loads is necessary for designing the product and efficiently utilizing the material. The macromechanical properties of a composite may be computed through a micromechanical analysis, and the design of a composite material may be started with knowledge of the constituent material properties. However, the validity of a micromechanical analysis is checked through experiments. Other purposes of mechanical characterization are to check the adequacy of the fabrication procedure to be adopted and to assure uniformity of the material.

1. *Hardness*—The hardness testing of plastics is most often measured by the Rockwell hardness test. It measures the resistance of the plastic toward indentation, thereby providing an empirical value of hardness. These hardness values do not necessarily correlate to other properties or fundamental characteristics. Rockwell hardness is generally chosen for "harder" plastics and is carried out as per ASTM D785 for plastics and composites. The samples are prepared in the form of small discs of rectangular geometry with a dimension of $50 \times 30 \times 3\ mm^3$.
2. *Tensile Strength and Modulus*—The failure of composites is a cumulative effect of a series of microscopic failures that include the failure of the matrix, the reinforcement, and interphases. In composites, fibres share the major load. Composites having uniaxially oriented fibres fail when the fibres suffer from fracture. In cases where the matrix cracks before the fibres fail, the ultimate strength of the composite is controlled by the fibre

alone. Therefore, the tensile strength of the composite is found as per Eq. 9.3.

$$\sigma_c = \sigma_f V_f \quad (9.3)$$

where:

- σ_c = Ultimate tensile strength of composite
- σ_f = Tensile strength of fibre
- V_f = Volume fraction of fibre

If the matrix does not fail before the fibres fail, the load is shared by the matrix until the fibres fail, and the ultimate tensile strength of composite is given by Eq. 9.4.

$$\sigma_c = \sigma_f V_f \left[\frac{E_m}{E_f} + (1 - V_f) \right] \quad (9.4)$$

where:

- σ_c = Ultimate tensile strength of composite
- σ_f = Tensile strength of fibre
- V_f = Volume fraction of fibre
- E_m = Tensile modulus of matrix
- E_f = Tensile modulus of fibre

In many applications such as pressure vessels and storage tanks, the utility of the composite is controlled by the matrix cracking and not by the strength of the reinforcement. The stress at which the matrix cracks is given by Eq. 9.5.

$$\sigma_c = \sigma_m V_f \left[\frac{E_m}{E_f} + (1 - V_f) \right] \quad (9.5)$$

where:

- σ_c = Ultimate tensile strength of composite
- σ_m = Tensile strength of matrix
- v_f = Volume fraction of fibre
- E_m = Tensile modulus of matrix
- E_f = Tensile modulus of fibre

In actual cases, the matrix fails much before the failure of the fibres mainly because of the stress concentration, fibre debonding, and strain magnification at microlevels. The tensile test is carried out as per ASTM D3039 on a Universal Testing Machine (UTM), as shown in Fig. 9.8.

This test provides the ultimate tensile strength, ultimate tensile strain, modulus of elasticity, and elongation at a break in the test direction.

3. *Flexural Strength and Modulus*—The flexural strength of a composite is dependent on the outer layers of the reinforcement. A well-cured monolithic composite will have higher flexural strength compared to its tensile strength. Flexural strength and modulus are tested as per ASTM 749 D. This 3-point bend test is shown in Fig. 9.9

Point to Ponder…

In a 3-point bend test, the upper portion of the sample is under compression, the bottom portion is under tension, and at the centre of the sample there is no load.

9

Fig. 9.8 Tensile testing of a composite on a UTM (Instron) [11]

The results are obtained in the form of flexural stress and strain. Flexural strength is the ability of the material to withstand bending forces applied perpendicular to its longitudinal axis. The stresses induced by the flexural load are a combination of compressive and tensile stresses. The calculations for flexural strength, strain, and modulus are given by Eqs. 9.6, 9.7, and 9.8, respectively.

$$\text{Flexural strength} = 3PL / 2bd^2 \tag{9.6}$$

$$\text{Flexural strain} = 6Dd / L^2 \tag{9.7}$$

$$\text{Flexural modulus} = \text{Flexural strength} / \text{Flexural strain} \tag{9.8}$$

where:

- P = Flexural load
- L = Span length
- b = Width
- d = Thickness
- D = Deflection

Fig. 9.9 Flexural ILSS test under 3-point bending [12]

4. *Compressive Strength*—Testing of the compressive strength of composites is a complex phenomenon—right from the fabrication of the test samples. The test samples used for compressive strength are cylindrical in shape. They have terminated internal plies that give rise to a large triangular resin pocket. The plies above this resin pocket are misaligned relative to the axial direction of the laminate. The test is carried out on a UTM as per ASTM D6484. This test method determines the open-hole compressive strength of multidirectional polymer matrix composite laminates [13].
5. *Shear Strength*—There are two types of shear strengths: in-plane shear strength (IPSS) and interlaminar shear strength (ILSS). The shear stress that distorts the lamina or layers in the composite is the *in-plane shear strength* (IPSS), which is the strength of the composite that resists distortion of lamina. The shear stress that acts on interlaminar planes of the composites is the *interlaminar shear stress* (ILSS). It measures the interlayer strength in a fabric-reinforced composite. ILSS is very low in comparison to IPSS. ILSS is an important material property for design because laminated composites are made of several plies and are bonded by polymeric materials. The ability of the material to resist shearing when one portion is moved from a stationary section is the shear strength.

   Shear strength is expressed as the maximum load required for shearing the specimen in such a manner that the moving portion has completely cleared the stationary portion. It is expressed in N/mm^2 based on the area of sheared edges. BS 4994 is the usual standard used for testing interlaminar shear strength. This test is also carried out as per ASTM D2344 on a

Universal Testing Machine (UTM), as shown in Fig. 9.8. The test method is similar to the flexural test method, except the thickness of the specimen is larger. The ratio of span/thickness of the test specimen is >16:1.

Impact Strength—Composites have good dampening properties. Long fibre-reinforced composites have good impact strength. The impact energy in composites is absorbed by the initiation of a number of cracks in the resin matrix on impact. Since the impact energy dissipates fast, the impact damages are localized.

Basically, impact strength is a measure of fracture toughness. High-impact materials are resistant to fracture. In short fibre composites, the impact resistance is improved if short fibres are directionally oriented and are set perpendicular to the impact force.

In randomly oriented fibre-reinforced composites, a good bond between the fibres and the resin matrix reduces the impact strength. This is contrary to long fibre-reinforced composites where a good bond between the fibre and the matrix imposes the impact strength. The poor bonding in randomly oriented short fibre-reinforced composites creates sites for absorbing shock rather transmitting it. This causes the absorption and diversifaction of impact energy to be fast, improving the impact strength of short fibre-reinforced composites. Notched impact strength is tested as per ASTM D-256 on a pendulum-type impact testing machine. In notched impact test, a notch is created at the centre of one longitudinal edge of the sample. The hammer is impacted just at notch during the test.

The *stress-intensity factor* (K) is used to determine the fracture toughness of most of the materials. There are three modes of fracture as illustrated in Fig. 9.10. The three modes of fracture are depicted by Roman numeral subscripts (I, II, and III) and are described here.

- *Mode I*: The crack plane is normal to the direction of a tensile force. It is the most common mode. It is denoted as K_{I}.
- *Mode II*: Shear force is applied on the crack plane due to a sliding motion. It is denoted as K_{II}.
- *Mode III*: Shear force is experienced due to a twisting motion at crack plane. It is denoted as K_{III}.

Fracture toughness is determined using ASTM E 399. The stress intensity factor is a function of loading, crack size, and structural geometry.

6. *Fatigue Strength*—The fatigue properties of a material represent its response to cyclic loading, which is a common occurrence in many practical applications. It is well recognized that the strength of a material is significantly reduced under cyclic loads. In the case of cyclic loading, the orientation of the fibres, selection of the matrix, type of loading (such as tension–tension, tension–compression, and compression–compression),

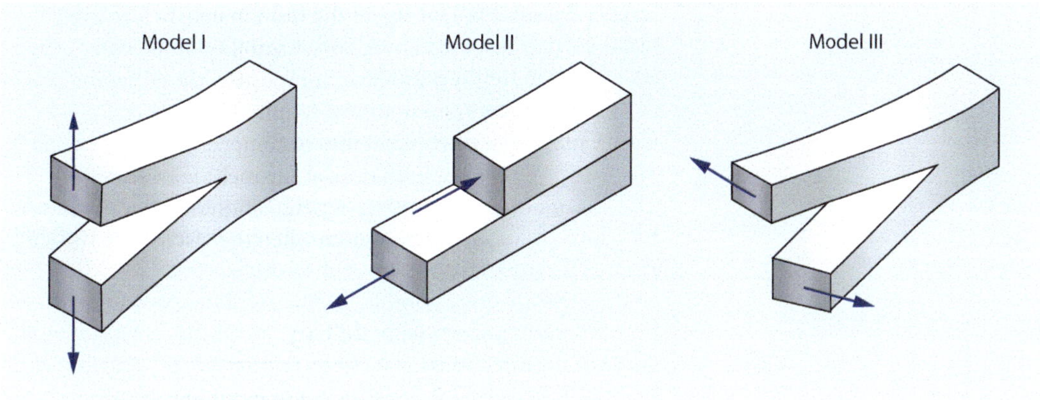

Fig. 9.10 Modes of fractures

frequency of loading, temperature, humidity, etc. all come into the picture. Heat build-up is also associated with cyclic loading (i.e. fatigue), and it also causes failure of the composite. *Fatigue failure* is generally understood as a cumulative effect of the gradual growth of internal flaws with a repeated or cyclic application of load. These failures are initiated in the matrix. Therefore, the matrix has a significant role in fatigue strength. The development of cracks due to fatigue causes a loss of modulus of the composite. Therefore, more load is borne by the fibres. Materials that are ductile in nature under normal operating conditions are known to fail in a brittle manner when they are subjected to repeated or cyclic stresses or strains. The cycle leading to failure depends on a number of variables, such as stress level, stress state, mode of cycling, frequency of load cycle, process history, material composition, and environmental conditions. The heat produced in the composite samples from fatigue is not conducted away due to the poor conductivity of composites. The heat build-up in composites softens the matrix, leading to debonding and causes a reduction in the fatigue strength. Due to all of these factors, a factor of safety of 2 is given to structures with a low cycle fatigue and of 4 for a high cycle fatigue.

The fatigue behaviour of a material is usually characterized by an S–N curve, which shows the relationship between the stress amplitude or maximum stress and the number of cycles leading to failure on a semi-logarithmic scale. The S–N curve of metals show an endurance limit. However, there is no such endurance limit in the case of composites. An S–N curve is obtained by testing a number of specimens at various stress levels under sinusoidal loading conditions. The number of cycles required for failure at a stress level is called the *fatigue life* of a material. There is a theoretical value for the stress amplitude below which the material will not fail for any number of cycles, which is called run out, fatigue limit, endurance limit, or fatigue strength. The fatigue strength of composite shown on S–N curve is the strength at which a sample does not fail for up to one million cycles. The fatigue test is carried out commonly in following three modes:

(i) Tension–tension mode
(ii) Tension–compression mode
(iii) Compression–compression mode

The interphase is also important in fatigue resistance; therefore, low-viscosity and high-strength resins score above highly viscous thermoplastic resins. For example, epoxy-based composites are quite good in fatigue resistance due to their high strength, low viscosity, and low shrinkage. A typical S–N curve of unidirectional glass fibre-reinforced epoxy composite in tension–tension mode (stress ratio $R = 0.1$) is shown in Fig. 9.11.

The majority of fatigue tests on fibre-reinforced composite materials have been performed with uniaxial tension–tension cycling. Figure 9.12 shows the stress–time diagram in a uniaxial tension–tension fatigue test. This test is carried out as per ASTM D3479–76. Figure 9.12 provides the details on stress range, mean stress value, maximum stress, minimum stress, stress amplitude, and frequency of stress application. At high cyclic frequencies, polymer matrix composites may generate appreciable heat due to the internal damping, which in turn increases the specimen temperature. Since a frequency-induced temperature rise can affect the fatigue performance adversely, low cyclic frequencies (less than 10 Hz) are preferred. In a stress-controlled test, the specimen is cycled between specified maximum and minimum loads so that constant load amplitude is maintained. In a strain-controlled test, the specimen is cycled between specified maximum and minimum strains so that constant strain amplitude is maintained. A unique feature of a fibre-reinforced composite material is that it exhibits a gradual softening or loss in stiffness due to the appearance of microscopic damages long before any visible

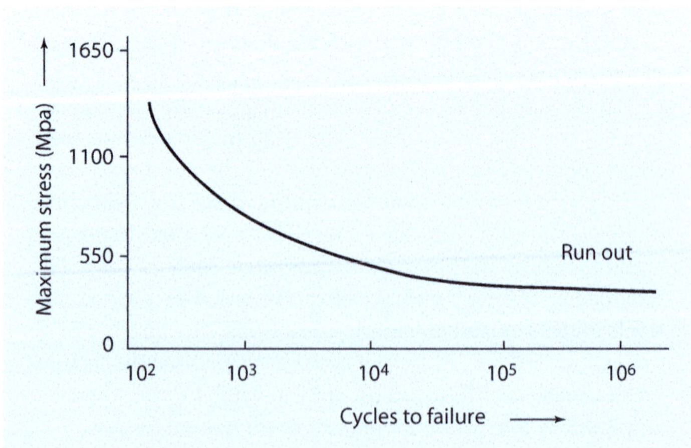

Fig. 9.11 Typical S–N curve of unidirectional glass fibre-reinforced epoxy composite in tension–tension mode

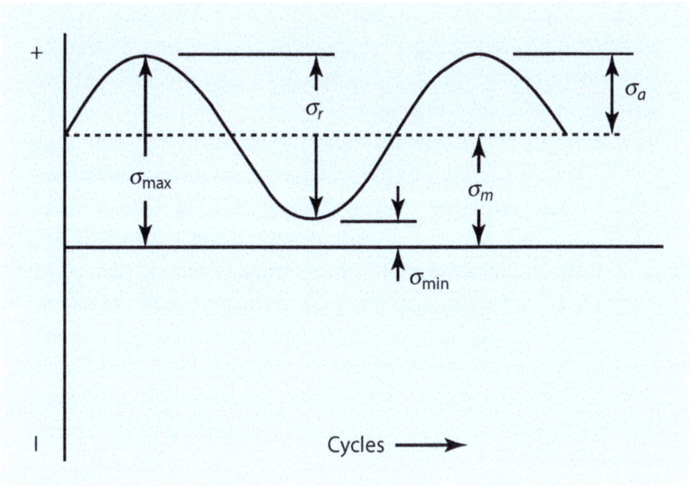

Fig. 9.12 Stress–time diagram in a fatigue test . σ_r = Stress range, σ_m = Mean stress, σ_a = Stress amplitude, σ_{max} = Maximum stress, σ_{min} = Minimum stress

damage occurs. As a result, the strain in the specimen increases in stress-controlled tests, but the stress in the specimen decreases in strain-controlled tests. In composites, fatigue failure occurs at the core first, and failure then propagates from the core to the surface. The reason of failure at the core is attributed to heat build-up at the core and the ability to to conduct it outside due to the insulating characteristic of polymeric composites. On the other hand, when the surface of the composite is in direct contact of air, heat is easily dissipated.

9.3.3 Electromagnetic (EM) Performance

The electromagnetic (EM) performance of a polymeric composite is evaluated based on its dielectric constant, loss tangent, transmission loss, and reflection loss. These are explained here.

1. *Dielectric Constant*—The *dielectric constant* (ε) is the capacity of charge storage by the insulating material on the application of an electric field. Low dielectric constant benefits good transmission of an EM wave because the loss of EM energy in polarization is at the minimum. The *relative dielectric constant* of the composite is calculated from Eq. 9.9.

$$\varepsilon'_{\mathrm{c}} = \frac{V_{\mathrm{m}} \log \varepsilon'_{\mathrm{f}} + V_{\mathrm{f}} \log \varepsilon'_{\mathrm{f}}}{V_{\mathrm{m}} + V_{\mathrm{f}}} \tag{9.9}$$

where:
- $\varepsilon'_{\mathrm{c}}$ = Relative dielectric constant of composite
- $\varepsilon'_{\mathrm{m}}$ = Relative dielectric constant of matrix
- $\varepsilon'_{\mathrm{f}}$ = Relative dielectric constant of fibre
- V_{m} = Volume fraction of matrix
- V_{f} = Volume fraction of fibre

Dielectric constant and transmission losses at various frequencies are measured using a *vector network analyzer* (VNA). A full two-port calibration is carried out using the TRL (*T*hru, *R*eflect, *L*ine) calibration method using calibration standards. The accuracy of the waveguide set-up needs to be in the range of 1%. The verification of the waveguide set-up is carried out with known standards. A typical sample size is $22.86 \times 10.16 \times 3.0$ mm^3. A micrometre (least count 0.01 mm) is used for the measurement of thickness, which is carried out on various locations (minimum 5) of the prepared sample in order to ensure the accuracy and uniformity of the thickness.

2. *Loss Tangent*—An electromagnetic wave has two components: electrical and magnetic. When an electrical field is applied to a dielectric material, electrical energy is absorbed in the polarization, such as electronic polarization, ionic polarization, dipole orientation, and space charge polarization. This polarization results in a loss of energy. Maximum loss of energy occurs when 63% of the total polarization process is completed and resonance occurs. This loss of energy is measured in the form of a *loss angle* (δ), and it is denoted as tan δ. The angle δ is the difference in phase angle—in other words, the current leads the voltage by an 90° angle-δ, which is called a loss angle (see Fig. 9.13). Otherwise, in general, current leads the voltage by 90°. The value of the loss tangent has to be a minimum for low transmission and reflection losses for an EM wave signal.

9

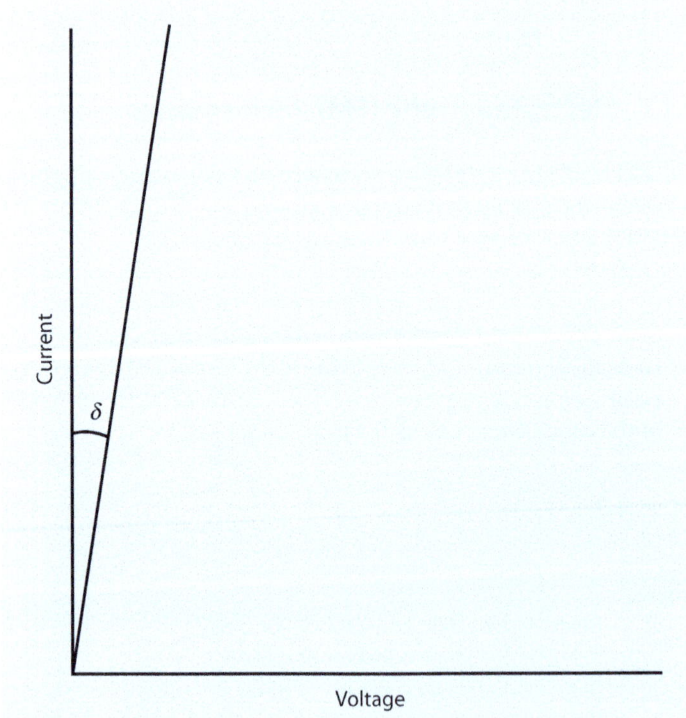

Fig. 9.13 Current leading the voltage by an angle 90°-δ

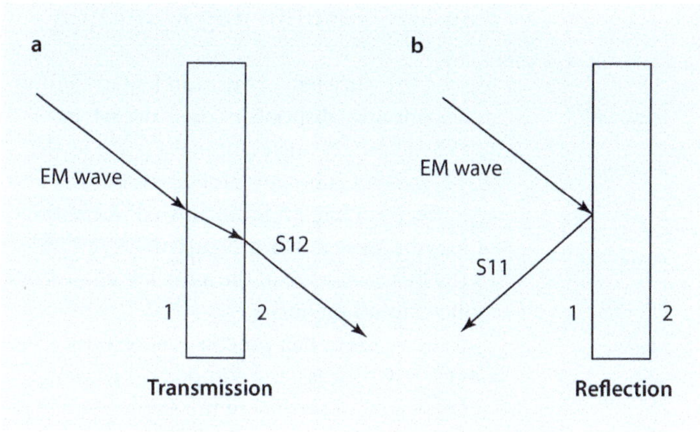

Fig. 9.14 **a** Transmission of an EM wave and **b** reflection of an EM wave through a dielectric material

The value of the loss tangent is experimentally calculated as the ratio of ε'' (imaginary part of permittivity) and ε' (real part of permittivity) of the sample, as given in Eq. 9.10.

$$\tan\delta = \varepsilon^{\wedge''} / \varepsilon^{\wedge'} \tag{9.10}$$

3. *Transmission Loss*—Transmission loss is a functional test of dielectric material that is used for radar-transparent structures. The loss of energy in transmission of electromagnetic wave through a dielectric medium is called *transmission loss*. In transmission loss, the EM wave travels through the thickness of dielectric material. The transmission of EM wave is denoted as S12 in Fig. 9.14a. Numerals 1 and 2 signify two surfaces (i.e. surface 1 and surface 2) of the dielectric material. S12 signifies that an incident EM wave falls on surface 1 and gets transmitted through surface 2. The losses occur due to the reason explained previously about the dielectric constant and loss tangent.
4. *Reflection Loss*—Reflection loss is also a functional test of dielectric material that is used for radar-transparent structures. In this, the loss of energy due to reflection from dielectric medium is measured. Reflection of an EM wave is denoted by S11 in Fig. 9.14b because an incident EM wave falls on surface 1 and gets reflected from itself without penetrating through the thickness of the sample. The losses occur due to the reason explained about dielectric constant and loss tangent.

Typical properties of reinforcements, matrices, and polymeric composites are tabulated in Table 9.1.

9.4 Advanced Test Methods for Composites

A number of NDT techniques are available for the assessment of a variety of features (such as V_f, voids, fibre wash, cracks, and layup order) in composites. Some of the important and advanced testing techniques are described next.

Table 9.1 Electrical properties of typical reinforcements, matrices, and polymeric composites

Materials	Dielectric constant (ε)	Loss tangent ($\tan\delta$)
Reinforcements		
E-glass fibre	6.0	0.004
Kevlar	4.1	0.02
Spectra	2.3	0.0004
Matrices		
Epoxy resin	3.6	0.04
Bismaleimide	3.3	0.004
Polyester resin	3.0	0.007
Phenolic resin	3.8	0.04
Composites		
E-glass/epoxy	4.4	0.02
E-glass/bismaleimide	4.7	0.02
E-glass/polyester	4.0	0.02
Kevlar/polyester	3.5	0.05
E-glass/phenolic	3.8	0.04

9.4.1 Radiography

Radiography provides a good overall understanding of polymeric composites. It detects vital parameters, such as general fibre orientation, fibrewash, kinks, wrinkles, and foreign objects. Translaminar cracks, which are aligned parallel to the direction of X-ray beam, are also easily detected in by radiography. The principle of industrial radiography that is employed in composites is shown in Fig. 9.15.

Translaminar cracks are present in composites in the form of thermal or shrinkage cracks. A commercially available X-ray generator with a low-voltage and beryllium window is used for radiography studies of composites. The voltage required for this X-ray apparatus is 10–15 KV, which is suitable for thin (less than 50 mm) sections of composites.

For the radiography of thick sections (50 mm and more) of composites to detect defects—including translaminar defects—a high-powered X-ray, such as a Lineac (*linear accelerator*), that works at a voltage of 15–50 KV is used. It requires many safety protocols because it poses fatal health hazards on exposure to human beings. X-ray radiography can be used to detect voids in composites, but only when they display absorption of a 2% incident X-ray or more. However, it is difficult to detect delaminations, volume

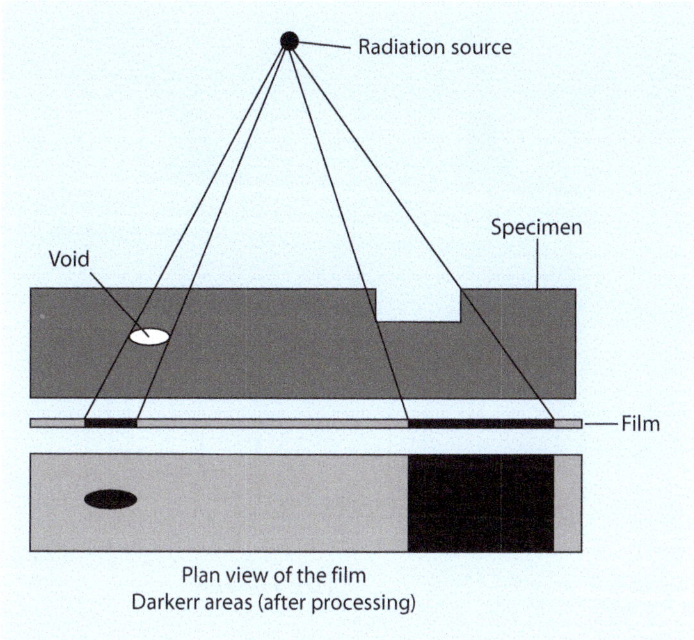

Fig. 9.15 Industrial radiography [14]

fraction (V_f), and cracks perpendicular to the beam using conventional radiography. Radiography is not suitable for the evaluation of the volume fraction (V_f) because it cannot provide accurate data on the mass or volume of the reinforcement. Also, radiography is not suitable for detection of interlaminar defects because the dimensions of these defects are very small (in microns) and the reinforcement fibres have high radiation absorption. X-ray microtomography is another promising NDT technique in which the X-ray beam focuses on 10-micron spot sizes and can detect interlaminar defects.

9.4.2 Ultrasonic Testing

There are two basic methods for ultrasonic testing: pulse echo and through-transmission tests. In the pulse-echo transmission method, the acoustic energy is transmitted through the solid section of the composite. An ultrasonic probe is placed on one surface or the outer surface of the composite. A good surface finish of the composite is required so that the probe makes good contact with the composite. If the acoustic energy encounters any flaw on its travel path in the form of delamination, voids, or air gaps, the ultrasonic signal is reflected back.

The principle of the *pulse-echo technique* is shown in Fig. 9.16. In this technique, an ultrasonic wave is transmitted through the material. A defect in the material reduces the amplitude of the reflected wave. The depth of the defect is determined by the ratio $D/E_{p.}$ In the presence of air or a void, the ultrasonic signal will be reflected back early without travelling through the full thickness of the composite. That's how the comparative assessment of the pres-

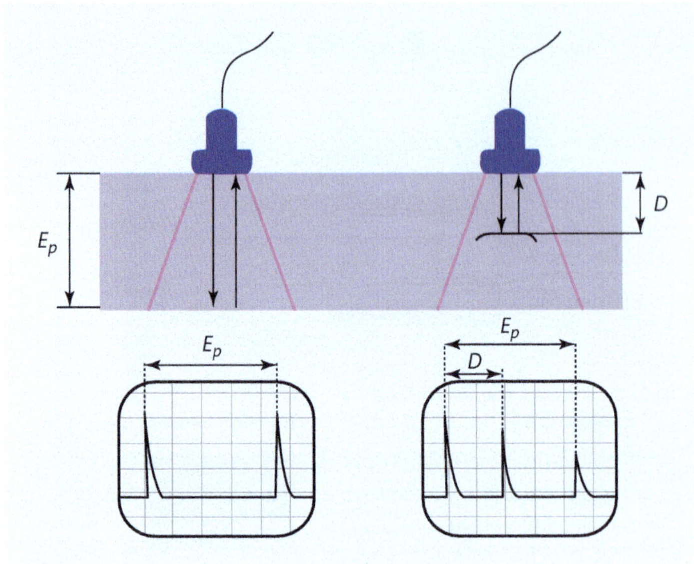

Fig. 9.16 Pulse-echo technique in field operation [17]

ence of a flaw is made. Pulse-echo transmission is carried out for detection of flaws in a thick section of composites where otherwise very high energy would be required using the through-transmission method [15, 16].

The pulse-echo technique is not found suitable for thin section composites due to the following limitations.

- In the cases of many defects, echoes of defects get superimposed and provide misleading results.
- Successive backwall echoes also superimpose and mislead the results.

The ultrasonic *through-transmission testing* technique has the advantage of not being affected by the surface finish, surface contour, etc. of composites. This technique is used for thin sections of composites. The acoustic energy is given in decibels (dB)—usually 20 dB—and is transmitted through the composite section and received on the opposite end. The ultrasonic through-transmission testing technique is used for the detection of flaws in composites and is carried out in the frequency range of 0.5–20 MHz. Ultrasonic attenuation provides the measure of the flaw present in the composite. It has been seen that ultrasonic attenuation is not very sensitive to the volume fraction (V_f). Interlaminar cracks affect ultrasonic attenuation in the same way as the voids present in composites. However, in the ultrasonic technique, it is difficult to differentiate between voids and interlaminar delamination. The most advanced ultrasonic NDT method is a C-scan method that is used for the detection of flaws in large and uniform composite products. The C-scan has syncronized scanning motions of the transmitter and the receiver in a water bath, which ensures transmission through liquid medium without any air. The ultrasonic signal attenuates in air; therefore, liquid medium (water) is required.

9.4.3 Eddy Current Testing

Eddy currents are circular electric currents induced—due to Faraday's law of induction—within conductors by a changing magnetic field in the conductor. *Eddy currents* flow in closed loops within conductors and in planes perpendicular to the magnetic field. Eddy current testing is possible for carbon fibre-reinforced composites (CFRC) due to the conductivity of carbon fibres. The inspection of a product using the eddy current method is shown in ◘ Fig. 9.17.

Eddy current testing is not possible for glass or Kevlar fibre-reinforced composites because they are insulators. Eddy current is an established inspection method for the detection of cracks, pits, inclusions, and voids on any metallic conductive surface, including ferrous and nonferrous in nature. This technique is capable of

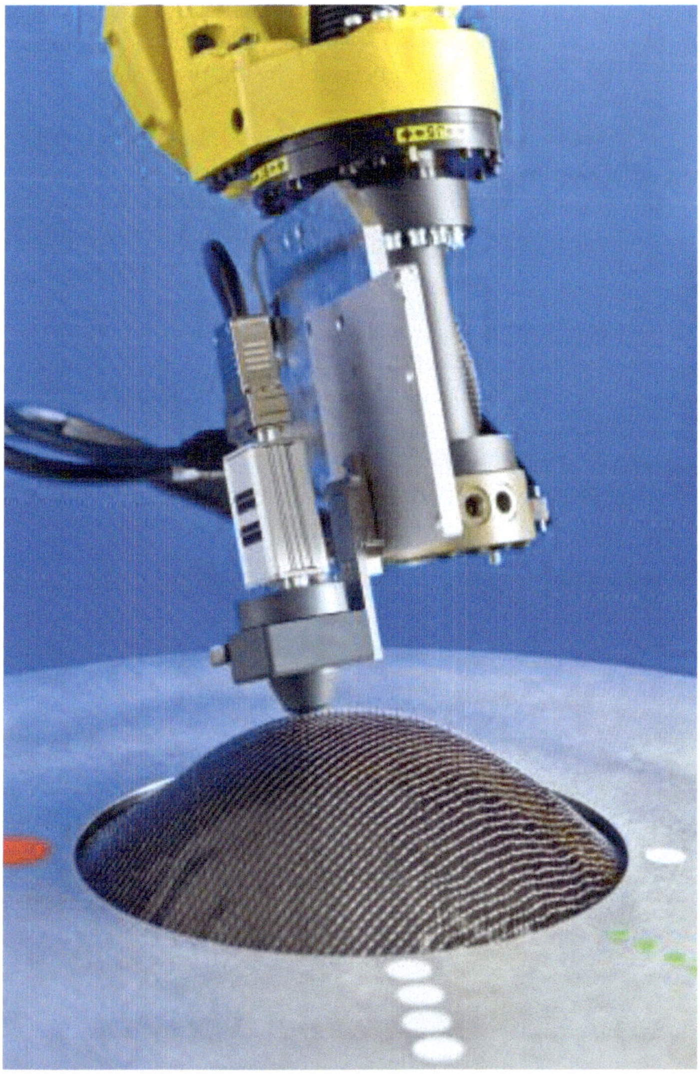

◘ **Fig. 9.17** Inspection by eddy current method [18]

9

finding flaws before they are visible and lead to creep or catastrophic failure [19–21]. Defects that are commonly detected using the eddy current technique are fatigue-induced cracking, thermal deterioration, vibration or mechanical damage, fretting, etc. Electrical currents are induced in carbon fibre-reinforced composites by the following two possible conduction paths when an eddy current probe is placed over them.

- *Resistive eddy current path* in which the induced eddy current passes from one fibre to another fibre at the point of fibre contact. At low frequencies, this conduction path becomes appropriate.
- *Resistive capacitive eddy current path* in which an interfibre capacitive path for an induced eddy current is present in addition to the resistive path. At high frequencies (>15 Hz), this conduction path becomes appropriate.

Changes in resistance in an eddy current probe can be used to determine the volume fraction (V_f). Eddy current techniques also may be used to detect cracks in CFRC and for the determination of the layup order in cross-plied CFRC.

9.4.4 Acoustic Emission Testing

Acoustic emissions may be defined as elastic waves spontaneously generated within the volume of the material due to the release of stored elastic energy as it undergoes plastic deformation, phase transformation, or fracture. Acoustic emission testing occurs during pressure tests of composite high-pressure cylinders, as shown in ◘ Fig. 9.18. When a force is applied on the composite structure, plastic deformation takes place at the tip of the flaw, leading to the generation of acoustic deformation. This characteristic is used for detecting flaws in composites using acoustic emission.

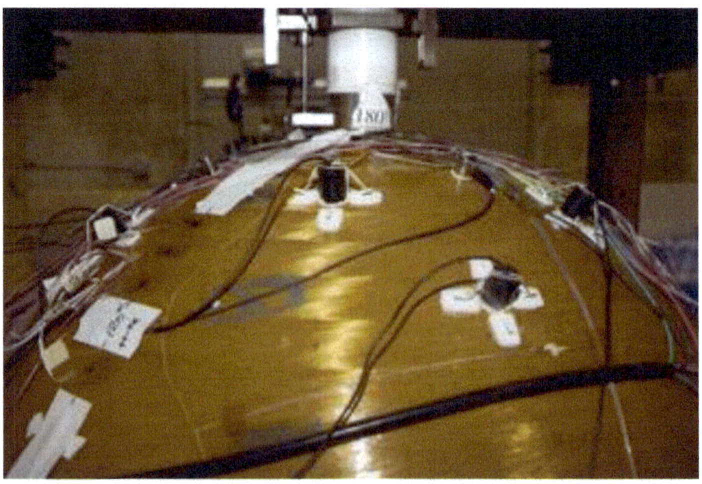

◘ **Fig. 9.18** Acoustic emission testing during pressure tests of composite high-pressure cylinders [22]

Acoustic Emission Testing is a test method for real-time measurement and on-line analysis of active or propagating damages in composites. It also gives insight into the governing micromechanical processes in the composite. For this detection and analysis, many transducers are required. With the help of computer, the data are analysed, and flaws in the composite are detected. This technique has been used for the detection of flaws in high-end composite applications, such as aerospace structures and pressure vessels. The acoustic emission technique also may be used for non-destructive evaluation of adhesive bonds because weaker bonds produce more emission at low stress levels compared to bonds that are stronger. It has been observed that fibre breakage causes an acoustic emission of higher magnitude compared to the emission generated either by the cracking of the matrix or debonding.

Point to Ponder...
The reliability of aerospace structures and pressure vessels needs to be very high due to safety considerations. At component and subsystem levels, the acoustic emission test is carried out for acceptance of these components and subsystems.

9.4.5 Shearography

Shearography is an optical NDT method for fast and accurate information about flaws or damages in composite structures. In this technique, a shearing interferometer detects extremely small (submicrometer) changes in surface deformation using a laser beam. When the composite is subjected under load, a proportional strain is induced on its surface. In this case, presence of defects or discontinuities present in the composite will cause the surface to deform unevenly. This deformation is interpreted through the interferometer as a shearographic image. Shearograpy provides real-time results and allows the measurement of large surfaces quickly and efficiently. Using the shearographic technique, first a reference image not under stress or strain is taken. Then the image with an induced strain due to stress is taken, and a comparison is made with the first image. The yielded result is the difference between the unloaded and loaded composites where the resulting surface deformations are directly linked to the out-of-plane surface strains of the material. The detection by shearography depends upon the strength and depth of the defects in the composite. Shearography is capable of detectingf most of the defects and discontinuities, including debonded materials, delaminations, cracks, a crushed core, wrinkling, fluid ingresses, voids, porosity, and damages beyond visual identification (BVIDs) in composites. Additional information on composites structures such as ply drops, bulkheads, overlaps, splices, stringers, and ribs also can be detected. The operation of a shearogrphy apparatus is shown in Fig. 9.19.

9.4.6 Vibrational NDT

The damage in polymeric composites may be manifested by the reduction in stiffness and increase in damping—irrespective of the location of the damage, where its distribution is in the form of microcracks. This vibration-based technique relies on models for identification of delamination in composites.

9

Fig. 9.19 Testing of a composite using shearography [23]

A change in the stiffness in the composite results in a decrease in the natural frequency of the composite. The measurement of the natural frequency of the composite at different stages of its life offers the possibility of detection and location of damage by comparing previous frequencies and their patterns to the latest frequencies and their patterns [24, 25]. The pattern of frequency in terms of amplitude changes with the type and severity of the damage in the composite.

9.4.7 Thermography

The increased use of composites in the aerospace industry, for automotives, and for other high-end applications has created a need for rapid surface-inspection techniques. Thermography is an important surface-inspection technique for composites. The thermographic image of the front portion of an aircraft is shown in Fig. 9.20. Subsurface failures such as delamination and debonding may occur in composites due to fatigue or impact force. *Thermography* is carried out using an infrared camera that identifies the purturbances of temperature on the structural surface caused due to the defects or damages in composites. The surface temperature differences may be caused by applying heat externally with an electric heater or hot air gun or by fatigue. Thus, hot spots are developed in the vicinity of stress concentrations caused by flaws. The intensity of hot spots is shown in the figure to be increasing with the colour code and numerical values assigned to them.

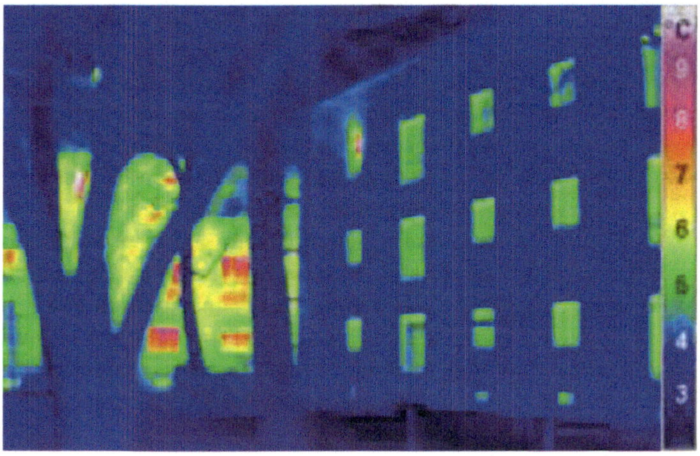

■ **Fig. 9.20** Thermographic image of a building [26]

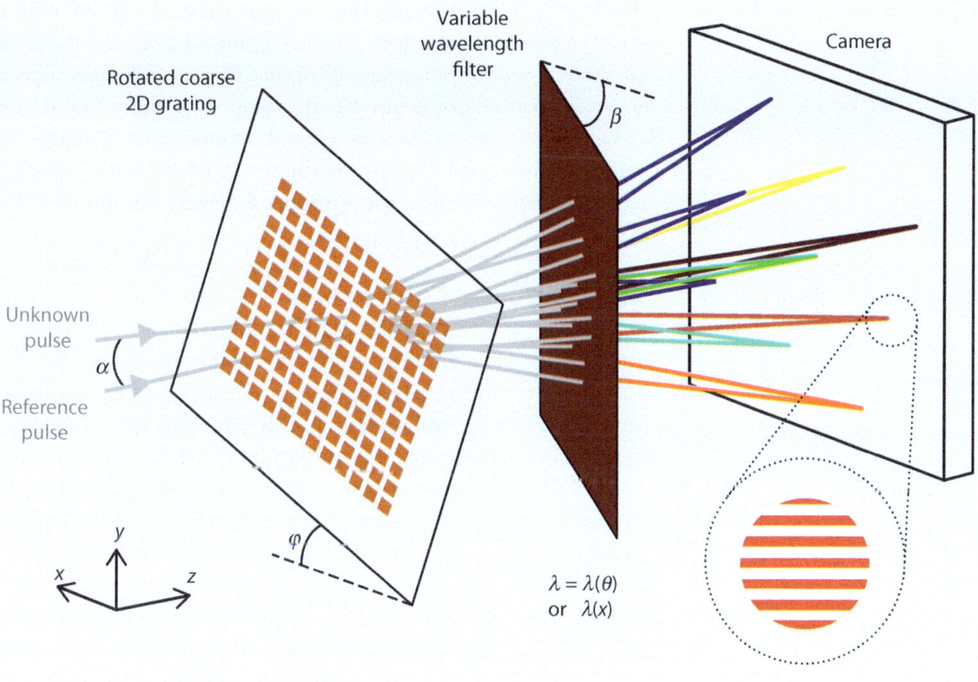

■ **Fig. 9.21** Schematic of holographic testing of composites [27]

9.4.8 Holography

In *holographic evaluation*, which is also referred to as holographic interferometry, the image (interferogram) of the composite is evalauted for deformation. Holography already has been proven to be an effective technique in the nondestructive testing of composites. A schematic of holographic testing of composites is shown in ■ Fig. 9.21. The principle of holographic evaluation is based on the local modification of the surface of the composite produced by a uniform stressing. These modifications are due to strain in the presence of local flaws (defects) under the surface of the composite. The visible deformation due to the generation of stresses occurs by illuminating the composites with laser light. All visual information on the illuminated composite surface is stored as a hologram on a photographic plate. The visibility of deformation on the composites provides information on flaws and defects.

One of the difficulties of holographic nondestructive testing is finding the method for stressing that is able to produce singularities in the strain state at the surface of composite being tested. This is a useful method for detecting local defects (such as flow marks, delaminations, voids, and broken reinforcement) in composites.

9.5 Structure–Property Relationship

Mechanical properties of polymer composites can be optimized by studying their structure–property relationships. Composites are characterized by the proportions of each phase, the shape of the reinforcement, and their arrangement in space. The phases of a composite often have different transport or related properties (i.e. thermal or electrical conductivity, gas permeability, modulus of elasticity). A given property of a composite is a function of the structure of the composite and the corresponding property of each component. Structure–property analytical relationships are interlinked [28].

9.5.1 Influence on Thermal Properties

The structure of a material influences its thermal properties greatly. The glass transition tempeature (T_g), melting point curing temperature, degradation temperature, etc. are influenced by the structure of the matrix, fibres, and their interphase. For example, an aromatic polyamide, such as Kevlar, has a very high melting point and high strength compared to an aliphatic polyimide, such as nylon 6 or nylon 6,6. In Kevlar, much of its thermal stability is provided by a rigid benzene ring. The melting point of Kevlar is 650 °C, while the melting point of nylon 6 is 240 °C. The chemical molecules in any material used for the reinforcement and the matrix are arranged in a special fashion, which is called their *stereochemistry*. The stereochemistry is responsible at a microlevel for any localized motion of molecules. The glass transition temperature (T_g) is highly dependent on the stereo arrangement of groups in the molecules. The temperature required for any translational movement of molecular segments is dependent upon how strongly the molecules in the polymers are arranged. For regular arrangements of the molecules, the heat energy required to initiate translational movement will be higher than the polymeric molecules that are not arranged in a regular fashion. For example, the well arranged PP molecules, which are crystalline and called isotactic PP, have a T_g of 0 °C, whereas atactic PP molecules, which have no regular arrangement, are completely amorphous and have a T_g of −20 °C. This behaviour of PP is shown schematically in ◘ Fig. 9.22. The figure shows that the translational movement at its T_g is low because every molecule in atactic PP is free to respond to the external heat stimuli. Whereas in isotactic PP, polymeric molecules are influence by the surrounding molecules and are not free to respond to heat stimuli [29].

In a similar fashion, the melting and thermal degradation temperatures are influenced. It is to be remembered that there are only secondary bonds among the polymeric chains of thermoplastic matrices. These bonds are broken at a temperature that may supply heat energy that is greater than the secondary bond energy. Once the external heat energy is more than the secondary bond energy among the polymeric molecules, the melting of the thermoplastic matrix takes place. When the heat energy supplied is in excess of the chemical bond energy of the polymeric molecules, the polymeric molecules themselves break. This is called the thermal degradation of thermoplastic matrices. This is shown schematically in ◘ Fig. 9.23.

In thermosetting composites, curing is a function of temperature or time or both. The glass transition tempeature (T_g) in thermosetting matrices is also dependent on its stereochemistry and cross-link density. Cross-link density has been discussed in detail in ► Chap. 5. The appropriate curing temperature, rate of curing, and curing time are responsible for uniform cross-link density of a matrix. The glass transition tempeature (T_g) in a thermosetting resin is caused by translational movement of chemical bond (i.e.

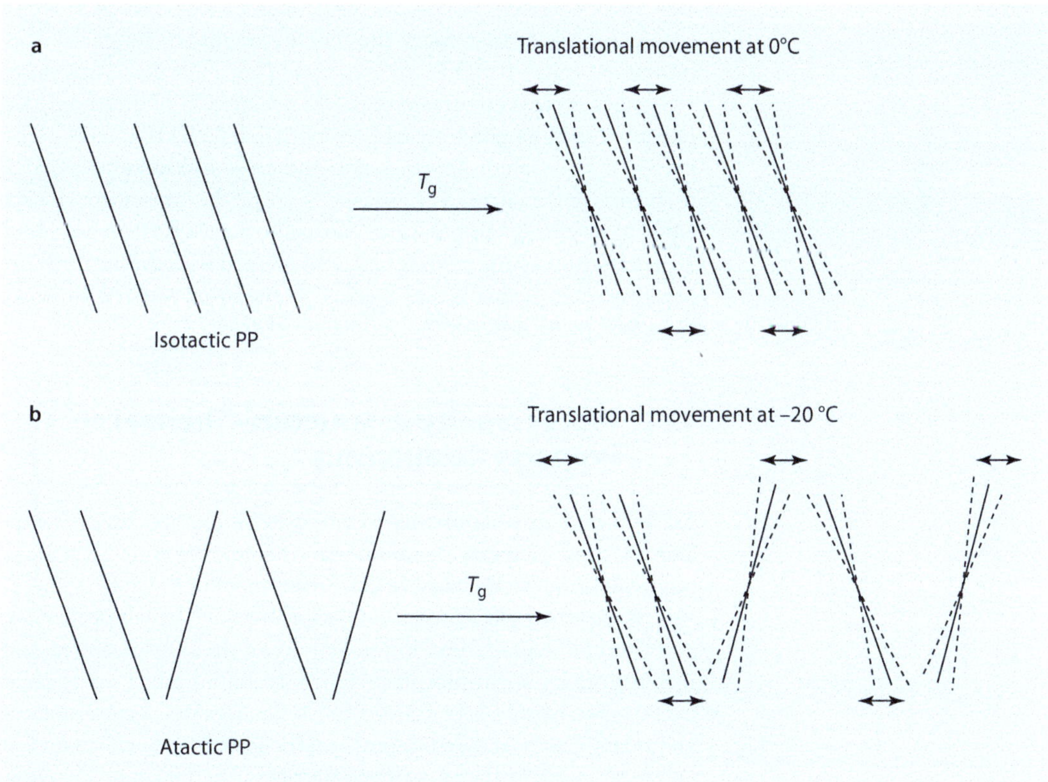

◘ **Fig. 9.22** Response of PP at T_g **a** isotactic PP and **b** atactic PP

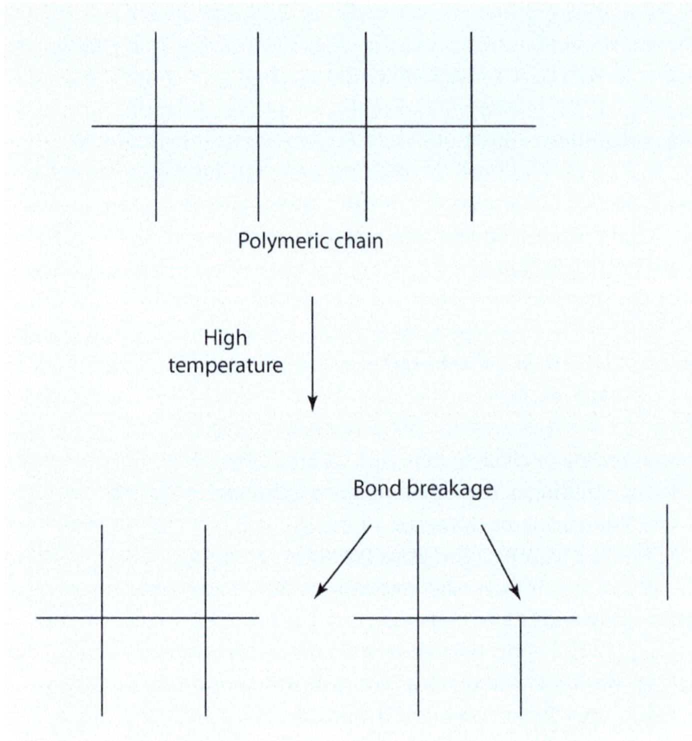

Fig. 9.23 Thermal degradation of polymeric chain at a high temperature

covalent bond); therefore, the energy required to cause this movement is much higher. For example, T_g of a thermosetting polyester such as vinyl ester resin is 180 °C, whereas T_g of thermoplastic polyester such as polybutylene terephthalate (PBT)) is 66 °C. The presence of covalent bonds makes the cross-linked network in a thermosetting matrix. This makes T_g of a thermosetting matrix equal to its curing temperature. This is the precisely the reason that T_g in a thermosetting resin is high and merged with the curing temperature. The mechanism of thermal degradation remains same, as discussed previously for thermoplastic matrices.

9.5.2 Processability in Structure–Thermal Property Relationship

The processability of polymeric composites influences the structure and thermal property relationship. The reinforcement causes a physical barrier to the translational movement of the matrix; therefore, it affects its glass transition tempeature (T_g) , melt properties, and thermal degradation. The reactivity of the reinforcement and matrices along with interphase has a profound effect on translational movement too. The interphase influences the performance of secondary bonds in thermoplastic matrices and the performance of covalent bonds in thermosetting matrices. The thermal behaviour of the matrix changes in the presence of reinforcement.

The smaller the size of the reinforcement, the more the physical barrier is present, affecting the translational movement of polymeric chains. Also, the reinforcement acts as a heat-soaking medium; therefore, it increases the performance of the composite at high temperatures. For example, in the ablation process, discussed earlier, the fibres provide a medium of heat soaking, and the matrix gets charred into a perforated mass like open cell foam. This cellular structure of charred matrix along with fibre reinforcement protects the composite structure from extreme heat in ablative composites.

9.5.3 Influence on Mechanical Properties

The mechanical properties are also greatly affected by the structure of composites. The structure of composites is guided by properties of reinforcement, matrices, and their interphases. For example, the rigidity and strength of Kevlar fibres are provided by the presence of a rigid benzene ring that is also responsible for the fibre's high melting temperature. Whereas nylon, which is aliphatic polyamide, has a lower strength and melting point compared to Kevlar. The factors of composite structure that influence mechanical properties are given here.

- Interphase
- Volume fractions
- Dimensions of reinforcement
- Uniformity of reinforcement
- Voids and air gaps
- Morphology
- Crystallization (for thermoplastic matrices)
- Cross-link density and its uniformity (for thermosetting matrices)

The prediction of performance depends upon the service environment. It has been always a challenge to predict the performance of polymeric composites. Generally, polymeric composites undergo catastrophic failures due to the brittleness of the material. Creep and stress relaxation are the main deformation mechanisms in polymeric composites.

Creep is a common phenomenon in polymers and also in polymeric composites. Creep is a permanent deformation under stress over a period of timethat is caused by the delayed response of the polymeric chains in polymers. The deformation stops once the polymeric chains reach a new equilibrium. Creep is also dependent on temperature. Polymeric composites creep faster at high temperatures.

Stress relaxation is another characteristic of polymeric materials and is a consequence of delayed molecular motions (as in creep) . Contrary to creep, stress relaxation is a reduction in the force required to maintain a constant deformation under constant strain. This is also a property of composite structures in which both reinforcement and matrix participate.

To Be Remembered

1. Characterization of any material establishes its character based on its morphology and molecular structure.
2. Sample collection and preparation are key to the testing of composites.
3. Smaller the size of the reinforcement, the more the physical barrier is presented in the translational movement of the polymeric chains.
4. Creep is a deformation under constant stress.
5. Stress relaxation is the reduction in stress on constant strain to the composite material.
6. Relation of morphology or internal structure of material with any property of a material is called a structure–property relationship.
7. DMA provides the viscoelastic response of composites.
8. Composites have viscoelastic behaviour due to the presence of a polymeric matrix.
9. Viscous parts of composite are responsible for dampening behaviour, whereas the elastic parts of composite give strength.
10. Fatigue strength and damage tolerance of composites are good.
11. Primary bond scission takes place in thermal degradation. Thermal degradation in composites is an irreversible process.

9

Exercises

1. Differentiate between DSC and DTA.
2. How will you measure the thermal stability of any composite?
3. How will you evaluate fibre volume fractions in carbon fibre-reinforced epoxy composite and glass fibre-reinforced PP composite?
4. What makes PEEK and PEI thermal resistant matrices?
5. What is the best process to evaluate the glass transition temperature?
6. Why is it difficult to detect the glass transition temperature in thermosetting composites?
7. Calculate the tensile strength and modulus of a glass fibre-reinforced polyester composite from the following load vs. deflection data.

Load (KN)	Deflection (mm)
2	0.2
5	0.5
8	0.8
11	1.1
15	1.6
19	2.0

8. Calculate the flexural strength and modulus of a glass fibre-reinforced PP composite in a 3-point bend test. The thickness, length, and width of the sample, as well as deflection and load are 3 mm, 60 mm, 10 mm, 3 mm, and 20 KN, respectively.
9. Explain the better mechanical performance of a carbon fibre-reinforced epoxy composite over a glass fibre-reinforced PP composite on the basis of their structure–property relationship.
10. What is the nondestructive technology used to estimate volume fractions?
11. Why are mechanical waves, such as ultrasound, used in NDT for detection of flaws in composite?
12. Why do you not use electromagnetic waves for detection of delamination or flaws in a composite?

References

1. Patel, G.N., Chance, R.R., Turi, E.A., and Khanna, Y.P., *Journal of American Chemical Society*, 100, 1978, p. 6644.
2. Seigmann, A. and Turi, E.A., *Journal of Macromolecule Science-Physics*, B10(4), 1974, p. 689.
3. Varma, I.K., Matrix resins for composites (seminar proceedings), "Development of resins moulding compounds and coupling agents for composite industry," *IIT Delhi*, New Delhi, March 12–14, 1984.
4. Shindé, S.L., and Goela, J.S., *High Thermal Conductivity Materials*, Springer, 2006.
5. Tsekmes, I.A., Kochetov, R., Morshuis, P.H.F., and Smit, J.J., *Thermal Conductivity of Polymeric Composites: A Review*, 2013 IEEE International Conference on Solid Dielectrics, Bologna, Italy, June 30–July 4, 2013.
6. "Recent trends in polymers and their composites," Proceedings of Seminar, *Department of Science and Technology*, New Delhi, December 1986.
7. Raghavan, V., *Materials Science and Engineering: A First Course*, Third Edition, Prentice Hall of India Private Ltd., 1989.
8. Rudge, A.W., Milne, K., Olver, A.D., and Knight, P., *The Handbook of Antenna Design (IEEE Electromagnetic Wave Series)*, London, Peter Peregrinus, 1986.
9. *Composite Material*, https://en.wikipedia.org/wiki/Composite_material.
10. Dwivedi, Mayank, Alam, Sarfaraz, Bhatnagar, Naresh, and Ghosh, Anup K., "Radar transparent glass fabric reinforced polyetherimide/Cloisite 30B nanocomposites," *Polymers for Advanced Technologies*, 22(12), 2011, pp. 2166–72.
11. *Tensile testing*, Wikipedia.
12. *Short Beam Shear*, Intertek Plastics Technology Laboratories, http://www.ptli.com/testlopedia/tests/Short_Beam_shear-D2344.asp.
13. Huang, Miinshiou, Jiang, Liang, Liaw, Peter K., Brooks, Charlie R., Seeley, Rodger, and Klarstrom, Dwaine L., "Using Acoustic Emission in Fatigue and Fracture Materials Research," *JOM*, November 1998 (vol. 50, no. 11) Retrieved 2011-12-05.
14. *Industrial Radiography*, https://en.wikipedia.org/wiki/Industrial_radiography.
15. *Aqualified*, http://aqualified.com/services/ultrasound-shearwave-ndt/.
16. "Mechanics of materials and acoustic emission testing," *BAM*, Federal institute for Materials research and Testing, http://www.bam.de/en/kompetenzen/fachabteilungen/abteilung_5/fg53/fb53_werkstoffmechanik_schallemissionspruefung.htm.
17. *Ultrasonic Testing*, https://en.wikipedia.org/wiki/Ultrasonic_testing
18. *Eddy current methods*, https://www.ikts.fraunhofer.de/en/departments/electronics_microsystems/systems_for_testing_analysis/eddy_current_methods.html.

19. *Thermography in Aerospace Industry*, Infratech GmbH, http://www.infratec.eu/thermography/application-area/process-optimisation/aviation.html.
20. Steeves, Craig A. and Fleck, Norman A., "Compressive strength of composite laminates with terminated internal plies," *Composites: Part A: Applied Science and Manufacturing*, 36, 2005, pp. 798–805.
21. ASTM D6484/D6484M – 14, *Standard Test Method for Open-Hole Compressive Strength of Polymer Matrix Composite Laminates*, http://www.astm.org/Standards/D6484.htm.
22. *Acousting Emission Testing*, NASA, https://www.nasa.gov/centers/wstf/supporting_capabilities/nondestructive_evaluation/acoustic_emissions_testing.html, 09 August 2016.
23. *Shearography*, https://en.wikipedia.org/wiki/Shearography.
24. Zou, Y., Tong, L., and Steven, G.P., "Vibration-Based Model-Dependant Damage (Delamination) Identification and Health Monitoring for Composite Structures—A Review." *Journal of Sound and Vibration*, v.2, 2000, pp. 357–378.
25. Cai, Jian, Qiu, Lei, Yuan, Shenfang, Shi, Lihua, Liu, PeiPei, and Liang, Dong, "Structural Health Monitoring for Composite Materials," *Book-Composites and Their Applications*, Edited by Ning Hu, August 22, 2012, Chapter 3.
26. Thermography, https://en.wikipedia.org/wiki/Thermography.
27. *Low-Tech – High Tech – and Really High Tech Non-Destructive Testing (for the technophiles)*, Holography, http://www.boatsurveyshawaii.com/Inspection_Methods.php.
28. Patro, S.K., and Upadayay, Paritosh, *Thesis on 'Failure characterization of FRP by scanning electron microscope technique'*, Department of metallurgical and Materials Engineering, NIT, Rourkela, 2008.
29. Charrier, J.-M., "Basic aspects of structure-property relationships for composites," *Polymer Engineering & Science*, 15(10), October 1975, pp. 731–746.

Correction to: Processability of Polymeric Composites

Correction to:

A. K. Ghosh, M. Dwivedi, Processability of Polymeric Composites, https://doi.org/10.1007/978-81-322-3933-8

The below listed late corrections have been carried out in the following chapters of the current version:

1. Chapter 1, p. 6, under heading Answer, symbol 'δ' was replaced by 'M'
2. Chapter 3, p. 72, line after equation (3.28) the recurrent terms "Sijkl" and "Cijkl" were deleted

The updated online versions of these chapters can be found at
https://doi.org/10.1007/978-81-322-3933-8_1
https://doi.org/10.1007/978-81-322-3933-8_3

A. K. Ghosh, M. Dwivedi, *Processability of Polymeric Composites*,
https://doi.org/10.1007/978-81-322-3933-8_10

Index

V

W

X

Y

Z

The manufacturer's authorised representative in the EU is Springer Nature Customer Service Centre GmbH, Europaplatz 3, 69115 Heidelberg, Germany. If you have any concerns regarding our products, please contact ProductSafety@springernature.com

Printed and bound by CPI Group (UK) Ltd, Croydon, CR0 4YY
12/08/2026
02194518-0004

Supplementery Information

A. K. Ghosh, M. Dwivedi, *Processability of Polymeric Composites*,
https://doi.org/10.1007/978-81-322-3933-8